THE AMBIDEXTROUS UNIVERSE

ALSO BY THE AUTHOR

THE INCREDIBLE DR. MATRIX

MARTIN GARDNER'S SIXTH BOOK OF MATHEMATICAL GAMES FROM SCIENTIFIC AMERICAN

MATHEMATICAL MAGIC SHOW

THE ANNOTATED ALICE

MARTIN GARDNER

The Ambidextrous Universe

MIRROR ASYMMETRY AND TIME-REVERSED WORLDS

Second revised, updated edition

ILLUSTRATED BY JOHN MACKEY

Charles Scribner's Sons, New York

LIBRARY OF CONGRESS CATALOGING IN PUBLICATION DATA

Gardner, Martin, 1914–
The ambidextrous universe.
Includes bibliographical references and index.
1. Parity nonconservation. 2. Symmetry (Physics)
I. Title.
QC173.G34 1978 501 78-16984
ISBN 0-684-15789-6 ISBN 0-684-15790-X pbk.

This book published simultaneously in the
United States of America and in Canada–

Printed in the United States of America

For my nephew
THEODORE GARDNER WEAVER

CONTENTS

THE AMBIDEXTROUS UNIVERSE

Preface to the First Edition

"The year 1957 was perhaps one of the most exciting years in the history of nuclear physics," writes D. Y. Bugg (reviewing a book on beta decay) in *The New Scientist*, August 16, 1962. "Early that year the news flashed from laboratory to laboratory that parity is not conserved. Professors waved their arms and talked excitedly about spin, about mirrors, and anti-worlds, and even undergraduates sensed that something remarkable was afoot."

The general public, too, sensed that something extraordinary had happened, especially when two Chinese-American physicists, Tsung Dao Lee and Chen Ning Yang, were awarded the Nobel Prize in physics for work that led to parity's downfall. But what is parity? How was it overthrown? Why are physicists so excited?

Fortunately, it is not necessary to know advanced mathematics or physics to understand the answers to these questions. It *is* necessary to grasp firmly the meaning of left–right symmetry and its curious role in the recent history of both the physical and biological sciences. In this book we begin with a deceptively simple question about mirrors. After examining the nature of mirror reversals in one, two, and three dimensions, followed by an interlude on left and right in magic and the fine arts, we plunge into a wide-ranging exploration of left–right symmetry and asymmetry in the natural world. This exploration culminates in an account of the fall of parity and an attempt to relate its fall to some of the deepest mysteries in modern physics.

In 1958 a small discovery in particle physics was reported at a meeting in Geneva. The discovery ironed out a theoretical difficulty that had long bothered Richard Feynman, a quantum-theory expert whom we will meet in chapter 22. "Dr. Feynman broke away from a food queue," the *New York Times* reported on September 5, "and danced a jig when he heard the news."

This book will not teach the reader any quantum theory. It will not even tell him why Dr. Feynman danced his jig. But it is the author's hope that the book's final chapters will convey to the general reader something of the jubilant, jig-dancing mood of the modern physicist when he shifts his attention from the macroworld of politics to the microworld of particles.

I would like to thank, without in any way implying responsibility for my biases and errors, Richard P. Feynman, who looked over an early draft of this book's manuscript and made numerous good suggestions, and Banesh Hoffmann, who straightened me out on several obscure points involved in one of the chapters.

For permission to reprint material, I am indebted to the following publishers and individuals: Alfred A. Knopf, Inc., for the poem "Cosmic Gall," copyright © 1960 by John Updike, from his book *Telephone Poles and Other Poems*; Bantam Books, Inc., for "The End," copyright © 1960 by Fredric Brown, from his book *Nightmares and Geezenstacks*; the *New Yorker* and Harold P. Furth for the poem "Perils of Modern Living," and the *New Yorker* and Edward Teller for Teller's letter commenting on Furth's poem, both poem and letter copyright © 1956 by The New Yorker Magazine, Inc.; and Michael Flanders for the lyrics of his song "Misalliance."

MARTIN GARDNER

Hastings-on-the-Hudson, New York
June 1964

PREFACE TO THE SECOND EDITION

In Carl Sandburg's *The People, Yes* there is an episode about a white man who draws a small circle in the sand and says, "This is what the Indian knows." He draws a larger circle around the small one and adds, "This is what the white man knows." The Indian takes the stick and draws an immense ring around both circles. "This is where the white man and the red man know nothing."

Let us vary the symbolism. The first circle is what the ancient Greeks knew. The second circle is what we know today. The third circle is what we will know a hundred years from now.

Scientific knowledge is like a continually expanding circle, but one that has no sharply defined perimeter. It is more like a cloud of beliefs with widely varying probabilities. At the center are convictions, such as the belief that the earth is round, that are "true" with a probability extremely close to 1. The farther from the center, the more likely a belief is not true. The cloud's boundary is therefore exceedingly fuzzy, made up of beliefs that are mere possibilities and about which experts disagree. In addition, portions of the cloud are forever shifting about, altering their distances from the center. A statement believed today with high probability tomorrow may be deemed false and banished to a region outside the cloud altogether. Other beliefs, now on the fringes, may drift nearer the center. Statements first made in crude form are constantly being elaborated and expressed with greater and greater precision.

Nobody knows if there is a limit to the cloud's size. There is even disagreement over what it means to say a belief is true, false, probably true, or probably false. There is no technique by which precise probability values can be assigned. Nevertheless, everyone agrees on one thing: the cloud expands. There is some sense in which one is justified in saying that all the science Aristotle knew is an extremely small part of what is known today.

This steady expansion of scientific knowledge is one of the few aspects of human history—perhaps the only aspect—about which we can say dogmatically that genuine progress takes place. Moreover, the progress itself progresses. The expansion occurs with steadily increasing rapidity. We have, for example, learned more in the last ten years about the structure of matter than was learned in the two thousand years that followed Aristotle.

For a science writer, this acceleration of scientific progress is a source of both delight and frustration: delight by the unending surprises, frustration by the quickness with which what one writes becomes obsolete. In 1952, intrigued by what in this book I call the Ozma problem, I wrote a paper, "Is Nature Ambidextrous?," that appeared in the *Journal of Philosophy and Phenomenological Research* (December 1952). In it I considered the possibility that someday a basic law of nature might prove to be left–right asymmetric. I ruled this out as almost unthinkable. Five years later, the unthinkable occurred. It was the shock of this discovery, the fall of parity, that prodded me into writing an entire book about mirror-reflection symmetry.

I completed this book, *The Ambidextrous Universe*, in 1963 and it was published the following year by Basic Books. No sooner was it on sale than another almost unthinkable event took place—time-invariance symmetry was found violated in the same weak interactions that violated parity. I was able to add a few pages about this to a Mentor paperback edition of the book, but it deserved much more than that. Indeed, it is the main reason for the edition you now hold.

Throughout, there are many small changes that correct and update, but the most important change is the addition of five new chapters on the latest speculations about time and how they bear on physics and cosmology. If I'm lucky, a few years may go by before the book is out of date again.

MARTIN GARDNER

1. MIRRORS

Some animals never seem to learn that mirror images are illusions. A parakeet, for example, is endlessly fascinated by what it sees in the reflecting toys placed inside its cage. It is hard to know what goes on within a bird's brain, but the parakeet's behavior suggests that it thinks it is seeing another bird. Dogs and cats are more intelligent. They lose interest in mirrors as soon as they learn that the images are not substantial. Chimpanzees also learn quickly that mirror images are illusory, but their high intelligence makes them intensely curious about what they are seeing. A chimp will play for hours with a pocket mirror. He makes faces at himself. He uses the mirror for looking at things in back of him. He will study the way an object looks when seen directly, then compare it with how the same object looks in a mirror.[1]

There is no better way to begin this book than by trying to see your image in the mirror with something like the wonder and curiosity of a chimpanzee. Imagine that one entire wall of a room is completely covered by a mirror. You are standing in front of this huge mirror, looking straight into it. Exactly what do you see?

Directly opposite you, of course, and staring straight back into your eyes, is a perfect image of yourself. Perfect? Not quite. Your face, like every face, is not exactly the same on its right and left sides. Perhaps you part your hair on the left. One ear or eyebrow may be a trifle higher than the other, your nose may twist slightly to one side, there may be a scar or birthmark on one cheek. If you look carefully enough you are sure to find some asymmetric features. When you do, you will note that on your mirror twin all these features are transposed. If you part your hair on the left, he parts his on the right, and similarly with all the other left-right features.

This reversal applies also, of course, to the room itself and all the objects in it. It is the *same* room, down to the last minute detail, yet at the same time curiously *different*. As Lewis Carroll's Alice said, when she peered into the mirror above the parlor mantel, everything in the room seems to "go the other way."

Well, not quite everything. Chairs look the same; so do most lamps and tables. If you hold a cup and saucer up to the mirror, it looks like a perfectly ordinary cup and saucer. But hold a clock up to a mirror and you see at once that it is changed. The numbers, instead of going "clockwise" around the dial, go "counterclockwise." (This reversal of clock faces, by the way, has provided important clues in many mystery novels. In A. E. W. Mason's famous murder mystery *The House of the Arrow*, a central clue is a girl's memory of what she saw on a clock face. It turns out that she had opened a door and glanced quickly at a clock without realizing she was seeing it in a mirror. Naturally, she misinterpreted what she saw.)

Hold a book up to a mirror. If you are far enough away from the reflection, the book appears unchanged. Move up close enough to read the title and you see immediately that the letters "go the other way." In fact, the words in reversed form are not easy to read. You may remember that, just after Alice had entered the looking-glass room, she opened a book on a table and came upon the world's greatest nonsense poem. This is how the first stanza was printed:

> Twas brillig, and the slithy toves
> Did gyre and gimble in the wabe:
> All mimsy were the borogoves,
> And the mome raths outgrabe.

Alice was clever enough to realize that if a mirror reflection is reflected, it is the same as if not reflected at all. "Why, it's a Looking-glass book, of course!" she exclaimed. "And, if I hold it up to a glass, the words will all go the right way again."

Small children are usually puzzled and delighted by this peculiar ability of mirrors to decode instantly a message written or printed with backward letters. Adults are no longer puzzled. They have become so accustomed to this property of mirrors that they take it for granted. They imagine they fully understand it. But do they really? Do *you* fully understand it?

Let me try to confuse you with a simple question. Why does a mirror reverse only the left and right sides of things, not up and down? Think this over carefully. The mirror's surface is perfectly smooth and flat. Its left and right sides do not differ in any way from its top and bottom portions. If it is capable of transposing the left side of your body to the right, and the right to the left, why doesn't it also switch your head and feet? Each line in the reversed stanza of "Jabberwocky" reads from right to left. Viewed in a mirror the lines read from left to right, but why does the top line remain on top, the bottom line on the bottom? Since the mirror exchanges left and right, what happens if we give the mirror a quarter turn clockwise? Will it turn the image of our face upside down? We know, of course, that no such thing will happen. Then why this spooky, persistent preference for left and right? Why does a mirror reverse the room horizontally but fail to turn it topsy-turvy?

I hope these questions are beginning to make you feel a bit more like an intelligent monkey contemplating his reflection in a pocket mirror. They are indeed puzzling questions. Try them on your friends. Chances are they will be just as puzzled. You will get plenty of embarrassed laughs and stammering attempts at explanation, but it will be surprising if anyone gives a clear, straightforward answer. With respect to mirrors, adults are more like cats and dogs than monkeys. They take mirror reflections for granted without attempting to get clear in their mind exactly what a mirror does.

To make matters even more bewildering, it is quite easy to construct mirrors that do not reverse left and right at all. For example, take two rectangular mirrors without frames and stand them on a table in the manner shown in Figure 1. The mirrors should be at right

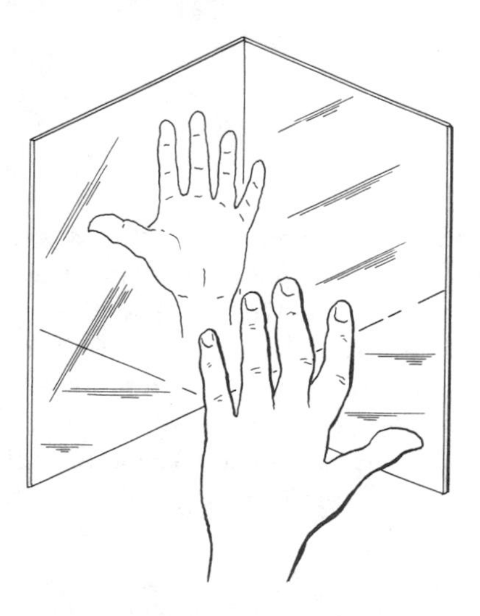

Figure 1. A double mirror that does not reverse images.

angles to each other, with two edges touching. Bend over and look directly into the mirrors. You will see an image of your face. If the image is too wide or too narrow, adjust the mirrors until it appears normal. But is it normal? Wink your right eye. Instead of your image's left eye winking—that is, the eye directly opposite your right eye—the image's *right* eye winks. The image is not a "normal" mirror image, but it is "normal" in the sense that it is a true, *unreversed* image. For the first time, you are seeing yourself in a mirror *exactly* as others see you.

Another way to make such a mirror is by bending a mirror—a sheet of thin metal, polished enough to give a mirror reflection—until it curves slightly as shown in Figure 2. When you obtain an undistorted

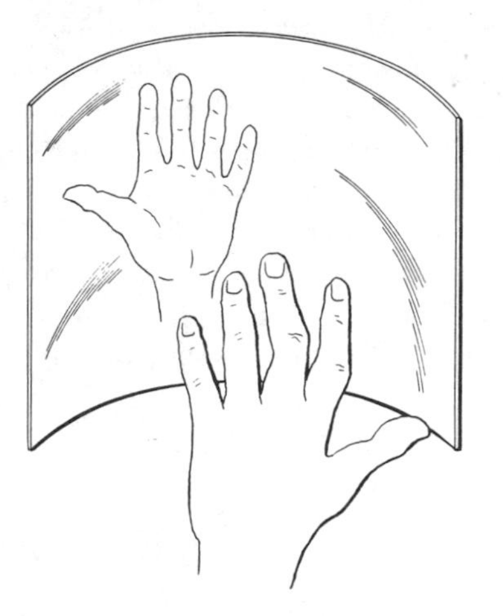

Figure 2. A single, curved mirror that does not reverse images.

image of your face, you will find it unreversed. You can test this easily by winking an eye or sticking your tongue out on one side. A curved mirror of this sort was known to the ancient Greeks. Plato, the famous Greek philosopher, describes it in his dialogue, *Timaeus*. It is also described by Lucretius, the Roman poet, in a section on mirrors in book 4 of his great science poem *On the Nature of Things*.

What happens to the image of your face if you take either of these strange mirrors and give it a quarter turn? The image promptly turns upside down! (See Figure 3.)

Apparently each mirror, when held a certain way, reverses neither right nor left, nor up and down. Held another way, the same mirror switches top and bottom.

Figure 3. Both mirrors, rotated 90 degrees, turn images upside down.

As a chimpanzee no doubt says to himself, while he reflects on mirror reflections, the matter deserves further study. We begin such a study in the next chapter by taking a closer look at exactly what a mirror does to geometrical figures in one and two dimensions. Before our study is finished, we will have explored many queer scientific truths, some frivolous, some not so frivolous. Two of the most stupendous scientific events of this century—the physicists' overthrow of parity and the biologists' discovery of the corkscrew structure of the molecule that carries the "genetic code"—are intimately connected with left and right and the nature of mirror reversals. In the end, our investigation will plunge us straight into some of the deepest, least-charted waters of contemporary science.

Notes

1. Recent research by Gordon G. Gallup, Jr. (which he reported in *American Psychologist*, May 1977) uses mirrors to show that chimps and orangutans have a self-consciousness not possessed by mammals outside the great-ape family. Chimps quickly learn that the image in a mirror is not another chimp. They use the mirror to groom parts of their body they cannot see, to pick food from their teeth, and so on. If a chimp is anesthetized, then painted around an ear with a bright red dye that is odorless and nonirritating, the awakened chimp is unaware of the painted spot until he sees it in a mirror. As soon as he sees it he tries to rub it off. Apes who have been taught to speak in signs will make a sign for themselves when asked who they see in a mirror. Self-awareness, like other human traits, seems clearly to be possessed in some degree by the great apes.

2. LINELAND AND FLATLAND

We live in a world of three dimensions, or, as the modern geometrician likes to say, a world of 3-space. Every solid object can be measured along a north–south axis, an east–west axis, an up–down axis. (A friend once told me that his college mathematics professor, a whimsical fellow, used to explain these three axes by running back and forth in front of his class, then running up and down the center aisle, and finally, hopping straight up and down!) Geometrical figures in 3-space are studied in solid geometry. If we confine our attention to two dimensions, we have plane geometry: the geometry of figures drawn on a flat, 2-space surface. We can go a step further down and consider figures of 1-space: one-dimensional figures that can be placed on a straight line. It is useful to consider the nature of mirror reflections in all three of these spaces.

Let us begin on the simplest level with Lineland, the space that consists of all the points along a single line that stretches off to infinity in both directions. Just for fun, imagine that this line is inhabited by a race of primitive creatures called Linelanders. Male Linelanders are long dashes with an eye (represented by a spot) at one end. Female Linelanders are shorter dashes, also with an eye at one end. The eyes do not develop until a Linelander becomes an adult. Children are simply short dashes without eyes. To make life more interesting for the Linelanders, we really should give them a world that consists of a complicated *network* of lines, so they can switch back and forth along the network and turn themselves around like freight cars on a railroad track, but this would unduly complicate matters, so we will keep them confined to a single line.

If a mirror is placed perpendicular to the line, as shown in Figure 4, we obtain a mirror reflection of the Linelanders. The picture shows

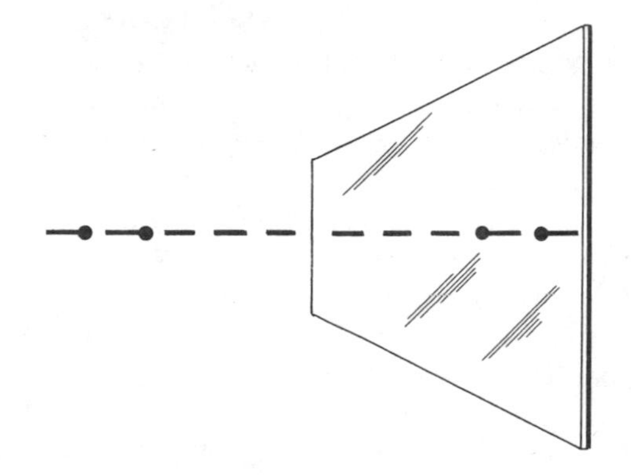

Figure 4. Linelanders and their mirror reflections.

the entire mirror, but as far as the Linelanders are concerned, their "mirror" is only a single point on their line. Note that a Linelander child is exactly like his mirror image. The mathematician puts it this way: The child is *superposable* on his mirror image. This means that we can imagine sliding the child along the line and into the mirror reflection, *without turning him around on the plane,* until he coincides, point for point, with his mirror twin. When this can be done, we say that the figure is symmetrical.

Are the adult Linelanders symmetric? No, because we cannot slide them into the looking-glass line and superpose them on their mirror images. This is because the ends of the adult Linelander are different. Suppose the line to run east and west. If an adult Linelander is facing east, his mirror image will face west. Of course we can turn him around and make him coincide point for point with his image, but to do this we have to remove him from the line and carry him through a higher dimension, the world of 2-space. Because the adult Linelander cannot be superposed on his mirror reflection, without entering a higher space, we say that his figure is asymmetric.

There is another way to distinguish between symmetry and asymmetry in Lineland. If a figure is symmetric, there is always a single point, exactly in the center of the figure, which divides the figure into identical halves, one a reflection of the other. Such a point is called the center of symmetry. If we place a mirror on this spot, perpendicular to the line and facing in either direction, the exposed half of the figure, together with the reflection, will reproduce the original figure. Would a Linelander with an eye at each end be symmetrical? Yes. Such a figure would also be superposable on its mirror image; there would be a center of symmetry dividing the figure into mirror-image halves.

Imagine a Lineland on which only three adults, A, B, and C, are living, all facing east. If we reverse one of them, say the middle one, this change will be instantly apparent to all three creatures. A and B are now looking at each other, B and C are now back to back. But if we reverse the entire line, that is to say, the entire "universe" of Lineland, the Linelanders themselves could not know that a change had taken place. In fact, it would be meaningless to them to say that any sort of change had occurred. *We* know that the line has been reversed, but that is because we live in 3-space and can see the universe of Lineland in relation to a world outside it. But the Linelanders cannot conceive of any dimensions higher than 1. They know only their own universe, the single line on which they live. As far as they are concerned, no change at all has occurred. Only when a *portion* of their universe is reversed can they become conscious of a change.

In Flatland, the 2-space world of plane geometry, things become more interesting, but with respect to mirror symmetry they remain essentially the same as before. In Figure 5 the artist has drawn a stylized conception of an asymmetric Flatlander and his reflection in a vertical mirror. (The mirror is shown in 3-space, but so far as the Flatlander is concerned, *his* mirror is no more than a straight line in

Figure 5. A Flatlander and his image in a vertical mirror.

front of him.) There is no way that he can be superposed on his mirror image; no way we can slide him around on the plane and make him coincide, point for point, with his reflection. If we could pick him up, like a paper doll, we could *turn him over* and put him back on the plane in reversed form. But this turning over would have to take place in 3-space. It cannot occur in the 2-space world of Flatland.

What happens if we hold the mirror above or below the Flatlander as shown in Figure 6? In this case, a top–bottom reversal occurs because it is the up– down axis that is perpendicular to the mirror. But the reversed image is really the same as before; it has merely changed its position on the plane. We can take either of the mirror images in Figure 6 and turn them so they coincide, point for point, with the mirror image in Figure 5. It does not matter in the least where we place the mirror; a reflection of an asymmetric Flatlander always produces the same reversed image.

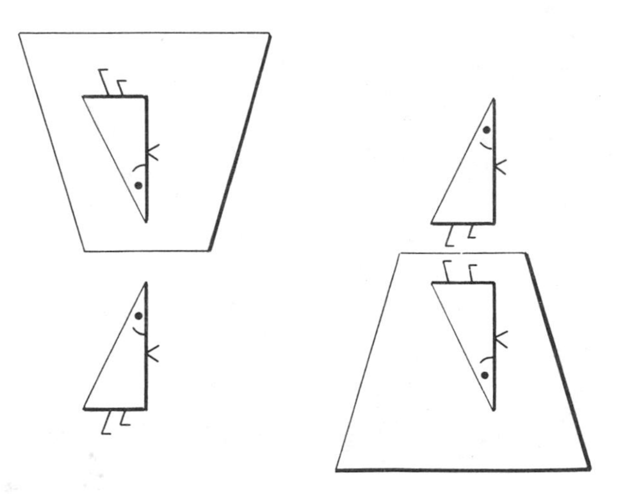

Figure 6. A Flatlander and his images in horizontal mirrors.

It is easy to draw all sorts of Flatland shapes that are symmetrical and therefore not reversed by the mirror. Squares, circles, ellipses, equilateral triangles, isosceles triangles, diamonds, hearts, spades, clubs—all are unchanged by reflection. In Lineland (as we learned) a symmetric figure possesses a point called the center of symmetry which divides it into mirror-image halves. In Flatland all symmetric figures can be bisected by a line called the axis of symmetry that does exactly the same thing. Figure 7 depicts a variety of symmetric plane figures. Axes of symmetry are shown as dotted lines. Note that the number of axes possessed by a figure may vary from one to infinity. The circle is the only plane figure that has an infinite number of them. Short of infinity, a figure can have any finite number of such axes. As in Lineland, if you place a mirror so its edge coincides with an axis of symmetry, the mirror reflection plus the exposed part of the figure will restore the shape of the original figure.

Any plane figure with *at least one* axis of symmetry is symmetric in the sense that it can be superposed, point for point, on its mirror image. Mathematicians talk about many other kinds of symmetry (some of them will be mentioned in chapter 11), but in this book we are concerned only with one kind: *reflection symmetry.* Whenever we speak of a figure as symmetric (regardless of the number of dimensions it has), we mean nothing more than the fact that it is identical with

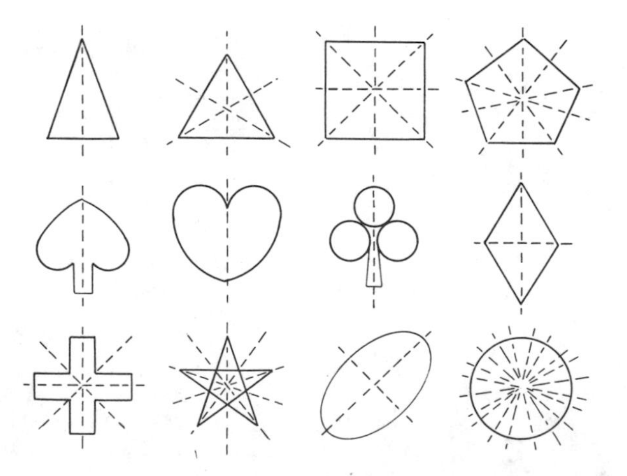

Figure 7. Plane figures with one or more axes of symmetry.

(superposable on) its mirror image without rotating it through a higher space. Whenever we speak of a figure as asymmetric we mean nothing more than the fact that it is not identical with (not superposable on) its mirror image.

It is easy to draw figures on the plane that are asymmetric. For example, the rhomboid, swastika, and spiral shown in Figure 8 cannot be superposed on their mirror images. If you try to bisect them down

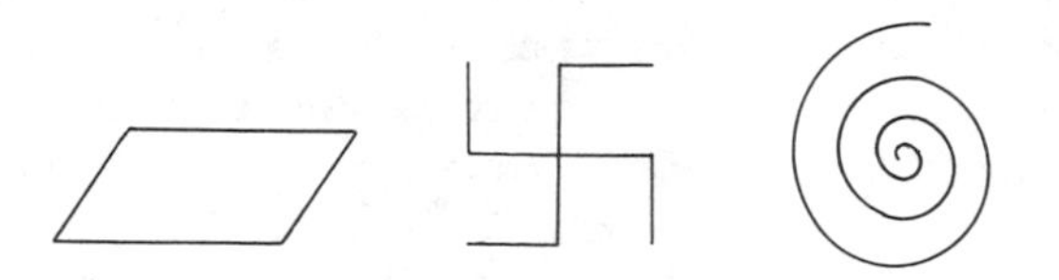

Figure 8. Asymmetric plane figures.

the middle with a line that divides them into mirror-image halves, you will find it impossible to do. There is no way to place the edge of a mirror on one of these figures so that the exposed part of the figure, together with its reflection, forms the original figure. For this reason, each figure can be drawn on the plane in two different forms. The swastika shown here is the form that was chosen by the Nazis for their symbol. Both forms of the swastika are ancient symbols that have been used by many different cultures.

Some of the capital letters in our alphabet are symmetric, some not. This suggests the first exercise for this book (all exercises are numbered, and answers are given in the back of the book):

A B C D E F G H I J K L M N O P
Q R S T U V W X Y Z

Figure 9. Which letters are symmetrical?

EXERCISE 1: *Which of the capital letters in Figure 9 are symmetric, which asymmetric?*

Try to answer the question without the help of a mirror. Remember, if a letter is symmetric it will be possible to find at least one line (maybe more) that will bisect it into mirror-image halves. If there is no such axis of symmetry the letter is asymmetric. Print all the symmetric letters on one sheet of paper and the asymmetric letters on another. Hold the sheet with the symmetric letters up to a mirror. If all of them are correct, it will be possible to turn the paper so that every letter will look the way it should. You may have to turn the paper one way to make one letter look right, another way to make another letter look right. This is because the axes of symmetry do not all run in the same direction. The letter *A*, for example, has a vertical axis of symmetry. It will look the same in a mirror when you hold the sheet right side up to the mirror. The letter *B*, however, has a horizontal axis of symmetry. It will seem at first to be reversed by the mirror, but turn the sheet upside down and the *B* will be normal looking again.

After you have verified all your symmetric letters in the mirror, see if you can draw on each letter all the axes of symmetry it possesses. You can do this on every letter except *O*. If the *O* had been formed as an ellipse, it would have only two axes, but because it is shown as a circle it has an infinite number.

Now hold up to the mirror the sheet on which you printed all the asymmetric letters. If all are correct, it will be impossible to turn the sheet so that any letter looks the way it should. All asymmetric letters have mirror images that "go the other way." Examine the letters on the sheet and you will see that it is impossible to bisect any of them with an axis of symmetry. These variations in the symmetry of letters make possible a number of amusing mirror tricks with words, but before explaining some of them (in chapter 4) we must devote a chapter to the symmetry and asymmetry of figures in 3-space, the solid three-dimensional world in which we live.

3. SOLIDLAND

In the world of 3-space, as in the worlds of 1-space and 2-space, all figures can be divided into two groups: those that are symmetrical and those that are asymmetrical. Symmetric solid figures are figures that can be superposed, point for point, on their mirror images. Asymmetric solid figures are those that cannot. Symmetric figures in 1-space (you will recall) have a *point* of symmetry; symmetric figures in 2-space have an axis, or *line*, of symmetry. As you might expect, symmetric figures in 3-space have what is called a *plane* of symmetry.

Some examples will make this clear. A sphere is a solid figure which obviously is identical with its mirror image. Just as a circle can be bisected by an infinite number of straight lines that divide it into mirror-image halves, so the sphere can be sliced through its center by an infinite number of planes that do the same thing. If a plane of symmetry is thought of as a mirror, then half the sphere plus its reflection in the mirror will restore the original sphere. Imagine a ping-pong ball cut in half. If you press the cut edges of either half-ball against a mirror, the reflection combines with the half-ball to restore the original ball.

The sphere is not the only solid figure that has an infinite number of planes of symmetry. A cylindrical cigarette, for example, has an infinity of such planes that pass through its axis, plus one plane of symmetry that cuts the center of the axis at right angles. An ice cream cone also has an infinity of planes of symmetry passing through its axis, but no plane of symmetry perpendicular to the axis. To be symmetrical a solid object must have at least one plane of symmetry, although it may have any finite number or an infinite number of them. The Great Pyramid of Egypt has four such planes. A brick has three. A table with a rectangular top has two. A chair and a coffee cup each have only one. Imagine a coffee cup sliced in half along its plane of symmetry. If we place either half against a mirror, the half and its reflection restore the original shape. (This of course is the meaning of a plane of symmetry.) The fact that a coffee cup has a plane of symmetry is what makes it a joke to speak of left- and right-handed cups.

Planes of symmetry have been sketched on all the solids in Figure

10 except the cube. Study the cube carefully and see if you can answer the following question:

EXERCISE 2: *How many planes of symmetry does the cube possess?*

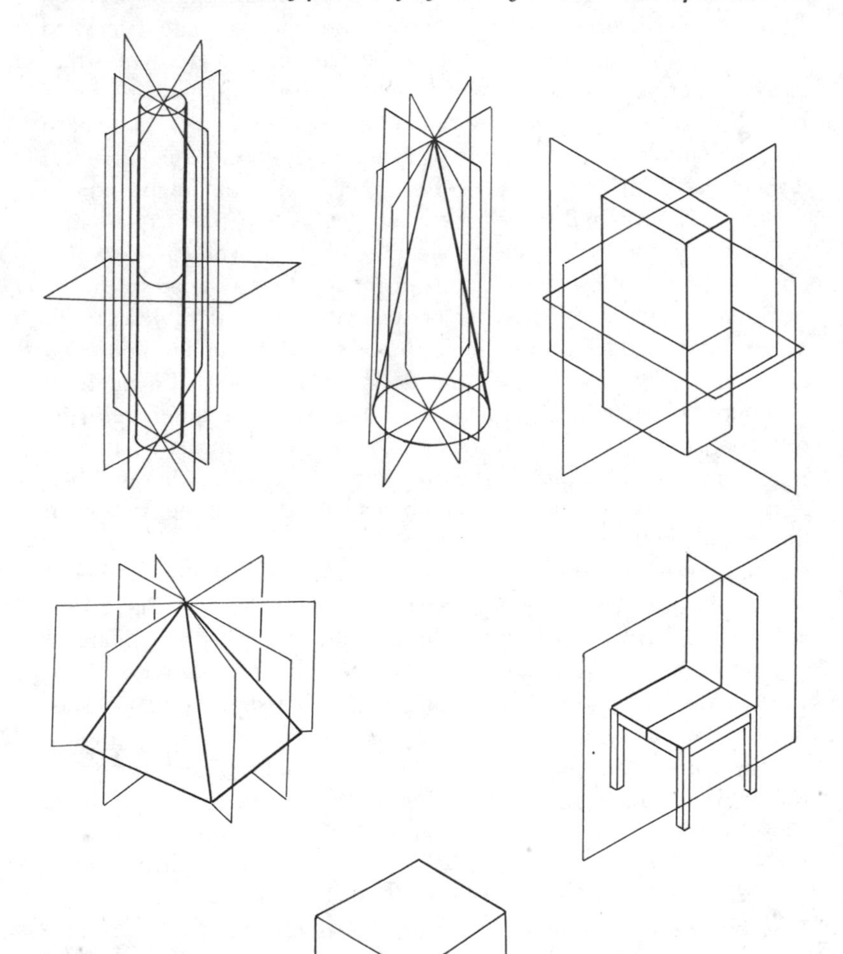

Figure 10. Planes of symmetry.

To superpose any symmetric object on its mirror image it may be necessary to turn one image around in 3-space to make it fit the other. Suppose, for instance, you are holding an ice cream cone up to a mirror. If you hold it as shown in Figure 11 (left), so that the mirror is parallel with one of the cone's planes of symmetry, you can superpose the two figures simply by moving them together until they coincide. But if you point the cone toward the mirror (Figure 11 right), the two figures are said to have a different orientation in 3-space. To make them coincide, you must turn one figure around until the two cones have the same orientation. The sphere never has to be turned because no matter how you hold it the mirror is always parallel with one of the sphere's infinite number of planes of symmetry.

Asymmetric solid objects are those that have no plane of symmetry, that can never be made to coincide with their reflections, no matter how they are turned. A simple example is provided by the helix, the curve of a spiral staircase and the red stripe on a candy cane. Just as the spiral is asymmetrical on the plane, so the helix, or three-dimensional spiral, is asymmetrical in 3-space. Try as you will, you will not be able to pass a plane through a helix in such a way that it cuts the helix into two mirror-image halves. Hold a helix up to a mirror. No matter how you turn it, in the mirror it always "goes the other way."

Every asymmetric solid figure has a mirror-image counterpart exactly like it in every respect except that it "goes the other way." Two asymmetric figures, each the mirror image of the other, are said to be

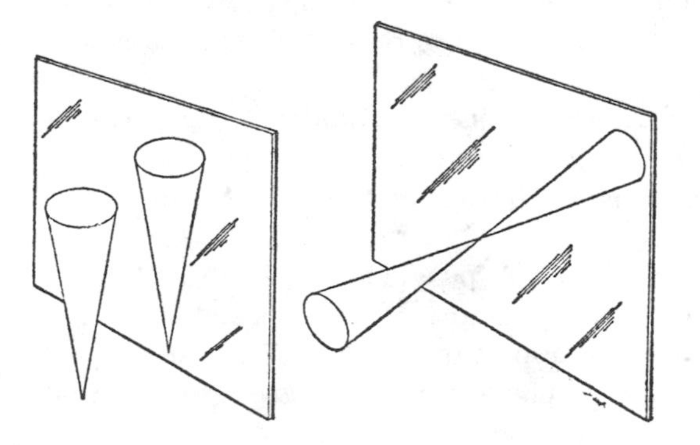

Figure 11. A cone in any orientation is superposable on its image.

enantiomorphs. Each is enantiomorphic to the other. A familiar example of a pair of enantiomorphs is your own pair of hands. Place them together, palm to palm. You will see that each is a mirror reflection of the other. This is such a commonplace example that enantiomorphs are often distinguished from each other by calling one left-handed and the other right-handed. A pair of gloves is a pair of enantiomorphs. Shoes provide another example. Your ears are enantiomorphs.

Any object with a helix on it is asymmetric: a corkscrew, or any type of screw, bolt, or nut that has helical threads. Screws are usually made so that they move forward by turning them clockwise with a screwdriver. Such screws are said to have right-handed threads. Sometimes left-handed screws are manufactured for special purposes. On some cars, for example, bolts and nuts that hold the wheels to the axles are right-handed on one side of the car, left-handed on the other. (This is because the turning of the wheels tends to rotate the nuts in a different direction on opposite sides of the car. The nuts have a "handedness" that keeps them from shaking loose while the car is moving.) Light bulbs that you buy in the store have right-handed helical threads around their base, but the bulbs formerly used in New York City subway cars were left-handed! This was to thwart thieves who otherwise might have stolen the bulbs to use at home. (Fluorescent lamps in special fixtures have replaced the bulbs.) Ever heard of a left-handed corkscrew? Yes, it can be bought in novelty shops as a practical joke. Hand it to someone when he wants to open a bottle and see how long it takes him to figure out why the thing doesn't work. Of course, if he turns it counterclockwise it works as well as a right-handed corkscrew.

EXERCISE 3: *Can you think of any reason why it is a universal convention throughout the world for screws and bolts (except those used for special purposes) to have right-handed threads?*

Look around you and you will be surprised at how many man-made objects are symmetrical at least in a general, overall way. In some cases, objects that appear to be symmetrical turn out not to be when examined more carefully. A pair of scissors, for example. The blades can cross each other in two different ways, one a reflection of the other. Most scissors are designed for use by right-handed people. If you are right-handed you know how awkward it is to hold the scissors in your left hand and trim your right fingernails. This awkwardness is due to more than just the fact that you are right-handed; it also arises from the fact that you are trying to use right-handed scissors in the wrong hand. To make them cut properly in your left hand you have to apply pressure on the handles in a most uncomfortable way. For this reason, special left-handed scissors are manufactured for left-handed tailors and other left-handers who have to use scissors constantly in their work.

Is a car symmetrical? In an overall way, yes, but when you consider such asymmetric features as the position of the steering wheel, then of course it isn't. The enantiomorph of an American car is a car with the steering wheel on the right, like the cars in England that drive on the left side of the road. Is a distant airplane, as you see it in the sky from the ground, symmetrical? Yes, except at night when asymmetry is introduced by the red light on the port (left) side and the green light on the starboard (right) side. Is an electric fan symmetrical? No, because its blades are parts of helicoid surfaces. If the blades were replaced with their enantiomorphs, the fan would blow the air backward instead of forward. The propellers of airplanes and ships are similarly asymmetrical. Is a piece of string symmetrical? Maybe. Examine it closely. If it is made of twisted strands, then of course it isn't symmetrical. Each strand forms a helix that will twist the opposite way when reflected.

EXERCISE 4: *Which of the following objects are asymmetrical?*

1. *Golf club*
2. *Fishing reel*
3. *Pliers*
4. *Wall can opener*
5. *Wall pencil sharpener*
6. *Salad fork*
7. *Sickle*
8. *Saxophone*
9. *Monkey wrench*
10. *Bowling ball*

That well-known topological curiosity, the Moebius strip, is asymmetrical. If you give a strip of paper a half-twist and paste together the ends, you obtain a surface with a single side and a single edge. But you can make the half-twist to the left or right. Twist it one way, you get a Moebius strip of one type. Twist it the other way, you get its enantiomorph, a strip of opposite handedness.

A simple overhand knot tied in a closed loop of rope also possesses handedness. Figure 12 shows an enantiomorphic pair of such knots.

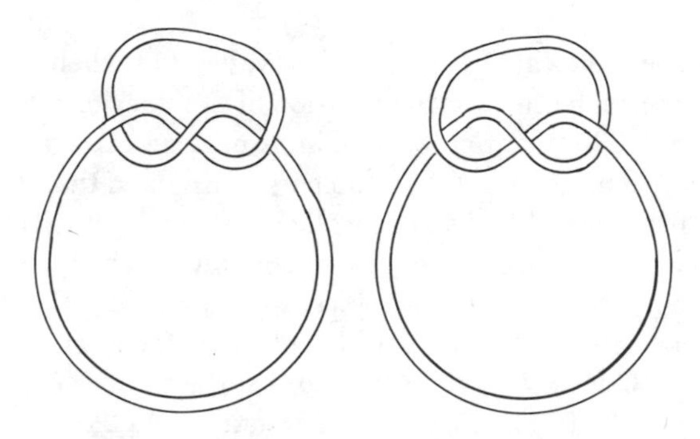

Figure 12. Right and left forms of an overhand knot.

No amount of manipulation of one knot can change it to its mirror-image twin. Has it ever occurred to you that when you fold your arms you are actually tying yourself into a simple overhand knot? The following experiment will make this clear. Place a piece of rope about three feet long on a table or have someone hold it for you. Fold your arms, then pick up an end of the rope in each hand. Unfold your arms. The knot in your arms will be transferred to the rope. The handedness of the knot will depend on how you folded your arms. Put aside the knot you have just tied and repeat the experiment with another piece of rope. This time, fold your arms the other way. The resulting knot will be the mirror image of the knot you tied before. If you stand in front of a mirror while you fold your arms and tie a knot in this manner, you will see your enantiomorph in the mirror fold his arms the other way and tie a knot of opposite handedness.

With this short introduction to reflection symmetry behind us, perhaps we can now answer that perplexing question raised in chapter

1: Why does a mirror reverse left and right but not up and down?

Curiously, the answer depends on the fact that our bodies, like the bodies of most animals, have only one plane of symmetry. It passes, of course, vertically through the center of the body, dividing the body into mirror-image halves. This is true only in a general, overall way. As mentioned in chapter 1, every face possesses minor asymmetries. Internally, of course, there are major asymmetries: heart on the left, appendix on the right, and so on. (In later chapters we will go into the asymmetry of living forms in some detail.) But superficially, animals and men have what biologists call "bilateral symmetry," meaning that the left side is a mirror reflection of the right. There is no resemblance between front and back, or between top and bottom. For this reason, and the fact that gravity pulls all things uniformly downward, we construct thousands of objects (chairs, tables, rooms, buildings, cars, trains, airplanes, and so on) that have (in the same superficial, overall way) bilateral symmetry. When we look into a mirror we see a duplicate of ourself, inside a room that duplicates the room in which we are standing. When we move our right hand, we see our twin move his left. We describe the reversal as a left–right one because it is the most convenient terminology for distinguishing a bilaterally symmetric figure from its enantiomorph. In a strict mathematical sense the mirror has not reversed left and right at all, it has reversed front and back.

To understand this, imagine yourself standing once more in front of that enormous mirror that covers one wall of the room. You are facing it directly, with your left side to the west, your right side to the east. Move your west hand. The hand on the west side of the mirror moves. Wink your east eye. The eye on the east side of the mirror winks. Your head is up, your feet down. Your image's head is up, his feet down. In other words, the east–west axis and the up–down axis keep their same orientation in 3-space. It is the front–back axis, the axis that runs north and south, perpendicular to the mirror, that has been reversed. You are facing north. Your twin faces south. Draw a north–south chalk line on the floor, perpendicular to the mirror, and label points along it, from north to south, in serial order: 1, 2, 3, up to 10. In the mirror, the points along the chalk line run from north to south in reverse order: 9, 8, 7, down to 1. In a strict mathematical sense the mirror has left unchanged the up–down axis and the east–west axis but has reversed the front–back axis. It is only because you imagine yourself standing behind the glass, facing the other way, that you speak of it as a left–right reversal.

This can be made clearer if you execute a "right-face" and stand facing east, your left side touching the mirror. As before, the mirror reverses only along the axis perpendicular to it. Because of the way you are standing, this is now in truth your left–right axis. *Now* you can say, in a strict geometrical sense, that the mirror has reversed your left and right sides, leaving unaltered your up–down and front–back axes.

Imagine a mirror on the ceiling or on the floor. Again, as always, the mirror reverses only the axis at right angles to its surface. This is now your up–down axis. The mirror leaves unaltered the positions in 3-space of your left and right sides, your front and back. It has, however, turned you upside down. But if you imagine yourself in the mirror, standing on your head, you see that when you move your left hand your upside-down twin moves his right hand. Even though the mirror has reversed only up and down, it is still convenient to you, because you are a bilaterally symmetric creature, to describe the mirror world by saying that left and right have been reversed. No matter how the mirror turns your world, you imagine yourself inside the turned world and you see that your left and right sides have been exchanged. You describe it as a left–right reversal rather than a front–back or a top–bottom reversal.

We can summarize it this way. A mirror, as you face it, shows absolutely no preference for left and right as against up and down. It *does* reverse the structure of a figure, point for point, along the axis perpendicular to the mirror. Such a reversal automatically changes an asymmetric figure to its enantiomorph. Because we ourselves are bilaterally symmetrical, we find it convenient to call this a left–right reversal. It is just a manner of speaking, a convention in the use of words.

The two trick mirrors described in chapter 1, the mirrors that give unreversed images, are mirrors that actually reverse figures along *two* axes! They reverse front and back, like an ordinary mirror, but (unlike an ordinary mirror) they also reverse your left and right sides. This double reversal, along two different axes, produces an image of the *same* handedness. You look into the mirror and note that when you wink your left eye, the reflected eye nearest the right side of the mirror winks. Because you think of yourself as standing inside the looking glass, facing the opposite direction, you say that the image winked its left eye, that no reversal has occurred.

After the trick mirror has been given a quarter turn, it continues to reverse the front–back axis, but now the other axis that it reverses is the up–down one. You see your face inverted. Inverted yes, but reversed no. If you imagine yourself inverted and inside the mirror, you see as before that when you wink your left eye he winks his left one also.

This may still seem confusing. You may have to read over the last seven paragraphs several times and think everything through carefully before you grasp exactly what an ordinary mirror does, or the two trick mirrors do, to asymmetric objects. To give your brain a rest before going on to more important matters, the next chapter will explain a number of amusing, easy-to-do tricks and stunts that are based on some of the ideas so far discussed.

Notes

1. Since this book was first published, three academic papers, each concerned with why a mirror reverses left–right but not up–down, have come to my attention. Jonathan Bennett, in "The Difference Between Right and Left" (*American Philosophical Quarterly*, vol. 7, July 1970), says that my explanation "is the only clear account" he knows. N. J. Block, in "Why Do Mirrors Reverse Right/Left but Not Up/Down?" (*Journal of Philosophy*, vol. 71, May 16, 1974; the title is printed in mirror-reflected form), thinks both of us are wrong. He distinguishes four different meanings of "reverse" and argues that in two of them mirrors do indeed reverse left and right but not up and down. Don Locke, writing on "Through the Looking Glass" (*Philosophical Review*, vol. 86, January 1977), contends that I am wholly wrong and Block only half right. Locke's article is by far the funniest of the three, even though he seems to be quite serious about it all.

Mathematicians find it hilarious that philosophers can still debate this hoary conundrum. Like the old puzzle of whether the hunter goes "around" the squirrel (see the second lecture of William James's

Pragmatism), the actual situation is so simple and easily understood that it requires a peculiar kind of mind to suppose that the mirror question is not trivial. The sole cause of confusion is, of course, the fact that the situation can be talked about in a variety of ways, using words and phrases that vary in meaning. Since the disagreements are entirely linguistic, the only way to resolve them is to seek agreement on the least confusing language to use, but now the question arises of "least confusing for whom?" A mathematician, a bartender, a philosopher? Most of the confusion stems from the fact that ordinary language defines left–right reversal in terms of our bilateral symmetry. This confusion vanishes in the more precise language of 3-space coordinate geometry, where there is no distinction between the coordinates except that they are called x, y, and z.

4. Magic

There are many stunts and magic tricks that illustrate in an entertaining way the principles of symmetry and asymmetry discussed in the previous chapters. One of the best of these tricks makes use of a package of Camel cigarettes.

Along one side of a Camel package you will find printed in large capital letters the words CHOICE QUALITY. Recalling the investigation of the symmetry of capital letters that you made in chapter 2, do you see what is remarkable about the word CHOICE? Not only is each letter symmetrical, but each has a horizontal axis of symmetry. For this reason the entire word has a horizontal axis of symmetry.[1] If you would hold the word to a mirror, *upside down*, the word's reflection in the mirror will appear unchanged. This is not the case with the companion word QUALITY. In QUALITY, as it appears on the cigarette package, only the letter *I* is symmetrical with respect to a horizontal axis. When held up to a mirror, no matter how the package is turned, the word's reflection is unreadable.

Magicians have a clever way of exploiting these facts in a mystifying parlor trick. Theodore H. Harwood, dean of the school of medicine at the University of North Dakota, has always been fond of this trick, and I am indebted to him for the following excellent manner of presenting it.

"Everyone knows," you say to your spectators, "that if a word is held up to a mirror, the mirror will reverse its letters. Most people do not know that cellophane has the same mirrorlike ability to reverse printing. Now if a reflected image is reflected a second time, it is the same as if it is not reflected at all. So if we allow a word to be reflected once by viewing it through cellophane, then we hold that reversed word up to a mirror, the word will be reflected a second time by the mirror and appear perfectly normal. Let me show you what I mean."

At this point you slide the cellophane wrapper down the sides of the Camel package until it covers CHOICE but leaves QUALITY exposed. Hold this side of the package up to a mirror, making sure you turn the package so the words are upside down. Figure 13 shows how the package appears in the mirror.

"Do you see?" you continue. "The word *choice*, having been reflected twice, once by the cellophane and once by the mirror, appears unchanged. But the word *quality*, since it is not reversed by the cellophane, appears reversed in the mirror just as you would expect."

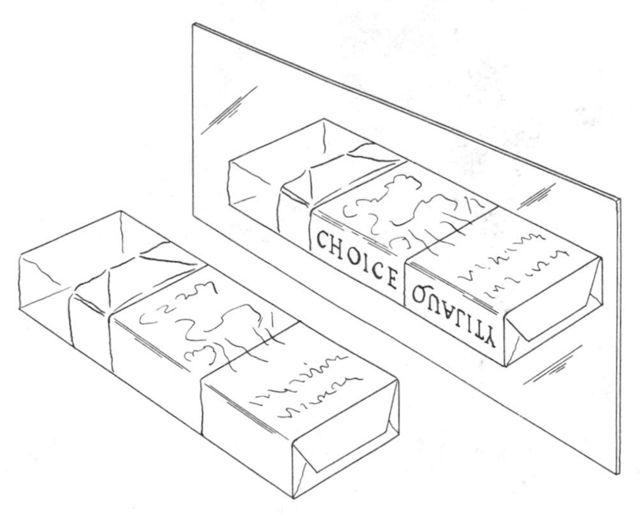

Figure 13. Why is *choice* reversed?

It is surprising how many people will buy this phony explanation. If someone insists on examining the pack, slide the cellophane back over both words and hand it to him. Chances are he'll move the cellophane back and forth over the words a good many times before the light finally breaks.[2]

Letters with vertical axes of symmetry remain unchanged when viewed right side up in a mirror. This explains why, if you hold Figure 14 up to a mirror, the boy's name is unchanged in the mirror, whereas the girl's name is reversed. You can tell your friends, when you show them this, that the mirror reverses black printing but not white.[3]

Figure 14. Why is *Timothy* not reversed?

It is possible to write numerals in such a way that they appear as letters in a mirror. Figure 15 seems to be an incorrect sum. View it in a mirror; you see it is correct after all.

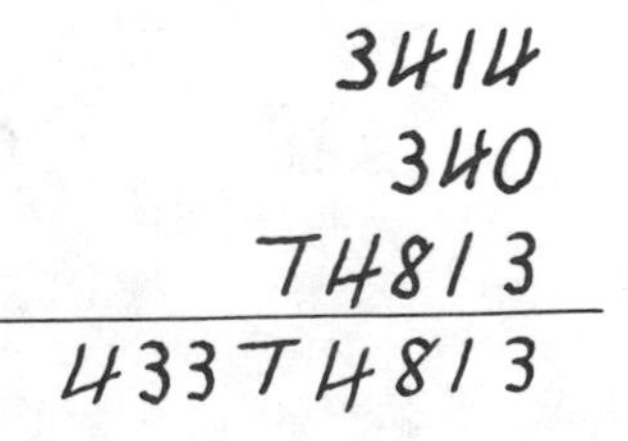

Figure 15. A mirror corrects this sum.

Many words change to other words when reflected right side up. For example, the word *bum* (with a lowercase *b*) changes to *mud* in the mirror. Cut these three letters out of paper (the larger the letters the better) and paste them on a wall or dresser mirror so they spell *bum*. Darken the room, then shine a flashlight on the letters. Shadows of the letters will be cast on the wall behind you.

EXERCISE 5: *If you turn your head and read the shadow letters on the wall, will the word be* bum *or* mud? *If you peer into the mirror and read the word on the wall in the room behind the looking glass, will it be* bum *or* mud? *Try to answer both questions before making an empirical test.*

The bilateral symmetry of the human face can be demonstrated by placing the edge of a pocket mirror (use a mirror with no frame) vertically down the center of a front-view photograph. The edge of the mirror rests, of course, on the picture's axis of symmetry. The exposed part of the face, together with the mirror image, forms the original face. However, owing to slight asymmetries in the features, the face will not look exactly the same.

Try this mirror test on front-view photographs of yourself and relatives and friends, or on front-view pictures of famous people that you find in magazines. It is sometimes amusing to see how different a face formed by two left sides of the picture (when the mirror faces left) will be from a composite face formed by two right sides (when the mirror faces right). Earlier in the century a group of German psychologists maintained that the two composite faces, seen in this way, represented the two basic sides of a person's personality. No reputable psychologist today takes this view seriously, but that need not prevent you from having the fun of "analyzing" friends by the mirror method. If you tilt the edge of the mirror slightly from the vertical you can transform even the best-looking face into a monstrosity.

In hotel lobbies and other buildings one often comes upon a square-shaped pillar, surrounded on all sides by mirrors. The bilateral symmetry of the human body makes possible a startling trick with such a pillar. Stand behind the pillar, with your nose pressed against the corner edge and half your body exposed to your viewers. The exposed half and its reflection form a composite image. (Shift a bit from side to side until your spectators tell you that the composite image looks normal.) Raise your exposed hand and pretend to blow on your finger. At the same time, lift your hat upward with your *concealed* hand. (Be sure to keep the hat horizontal.) It looks exactly as if your hat suddenly blew high into the air. Take the finger from your mouth and allow the hat to settle slowly back on your head. Many people are utterly mystified by this simple stunt.

For an encore, lift your exposed leg. Both legs of the composite image will go up like a jumping jack on a string. At the same time, roll your eyes rapidly around in circles. To your viewers, one eye revolves clockwise, the other counterclockwise.

If you place the edge of a mirror on any type of figure or pattern, a composite picture with bilateral symmetry results. Perhaps you made inkblot pictures as a child. Simply let a few blobs of ink fall on a sheet of paper, fold the paper in half so the crease passes through the ink, and press the two halves together. When you open the paper you will have a bilaterally symmetrical design. The well-known Rorschach test, used by psychiatrists as a diagnostic aid, makes use of inkblot pictures that were originally formed in this way. The crease in the paper marks, of course, the pattern's axis of symmetry.

If two mirrors are formed into a V and placed on a figure or pattern, the result is a series of reflections. By adjusting the angle until it is a submultiple of 180 degrees, pleasing patterns with an even number of axes of symmetry can be formed. If the angle is $180 \div 2 = 90$ degrees, the pattern will have two axes of symmetry, not quite enough to be interesting. An angle of $180 \div 3 = 60$ degrees produces the striking hexagonal patterns of snowflakes, with three axes of symmetry. Note that the reflections alternate in handedness as you go around the pattern. Place the mirrors at a 60-degree angle on the colored illustration of a magazine, then slide them slowly around the page, keeping them always at the same angle. The abstract hexagonal patterns will change rhythmically to different designs, preserving at all times their beautiful hexagonal symmetry. Most kaleidoscopes are made with mirrors at 60-degree angles to form similar designs by reflecting random patterns created by bits of colored glass.

There is a novel form of kaleidoscope called a teleidoscope. Instead of containing bits of colored glass, it has a convex lens at each end.

This makes it a telescope as well. Any scene viewed through the teleidoscope is reflected seven times by two mirrors placed at an angle of 180 ÷ 4 = 45 degrees. In this case the pattern is octagonal, with four axes of symmetry.

A puzzling stunt involving left and right structures can be presented with two or more ordinary dice. If you stack three dice as shown in Figure 16 and cover the stack with a half-dollar, four sides of each die will be visible (as you walk around the stack) and two sides of each die

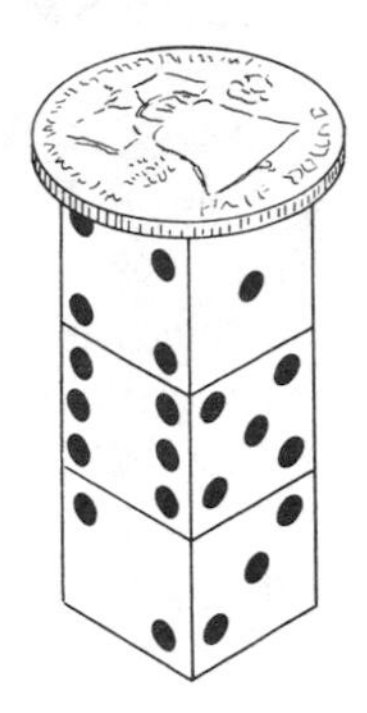

Figure 16. Can you name the top of each die?

will be concealed. Can you guess correctly the number that shows on the top surface of each die in the illustration? Opposite sides of a die must total 7, so it is easy to determine that the top of the bottom die must be either 6 or 1. Similarly, the top of the middle die must be either 4 or 3, and the top of the top die must be either 5 or 2. How can you tell which of each pair of digits is the correct one?

The method lies in the fact that there are only two ways that the sides of dice can be numbered, provided opposite sides sum to 7. One way is a mirror image of the other. If you hold a die as shown in Figure 17, with the 1, 2, 3 faces toward you and the 1 on top, you will see that the digits, in serial order, go counterclockwise as indicated by the

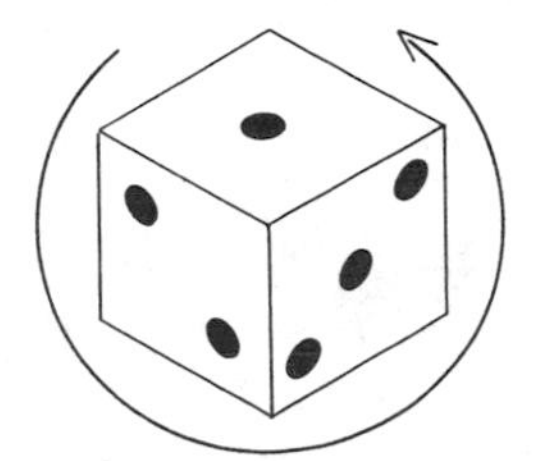

Figure 17. All modern dice are left-handed.

arrow. All modern dice are made this way. In past ages, dice were sometimes made one way, sometimes the other. Cubical dice that "seven-up" all around go back as far as ancient Egypt, where they were made in both right- and left-handed forms.

Once you know that all modern dice are "left-handed," it is not difficult to name the top faces in a stack of dice topped with a coin. Simply look at two faces of each die and visualize in your mind where the 1, 2, 3 sides must be. With a little practice, bearing in mind that opposite sides total 7 and that the 1, 2, 3 go counterclockwise, this is not at all hard to do.

EXERCISE 6: *Name the top faces of each die in Figure 16.*

Not one person in a thousand is able to guess the tops of dice correctly when they are stacked in this manner. I have seen gamblers perform this trick at casinos, with six or more dice that someone stacks at random while the gambler's back is turned. The gambler takes one glance at the stack and instantly calls out the uppermost faces while the numbers are verified one at a time by removing dice from the top of the stack. The feat seldom fails to make an impression and usually sparks a confused discussion of how the spots on dice are arranged.

If you try these tricks on your friends you'll find them entertaining, with a mathematical flavor that heightens their interest. But we have more important things to look into. The next chapter will discuss the role of reflection symmetry in painting and, what is more surprising, its role in music and poetry.

Notes

1. After this chapter was written, Richard Welling, a graphic designer in Hartford, Connecticut, sent me a proof of an advertisement picturing a cookbook put out by General Foods Kitchens. The word COOKBOOK appears in large type near the center of the book's cover. The picture had inadvertently been reversed and turned upside down, but because of the horizontal axis of symmetry through the

word COOKBOOK, no one had noticed the mistake. Luckily Mr. Welling caught the error before the advertisement was printed.

2. A similar trick is to print the words CARBON DIOXIDE, using a different color for each word. When they are held upside down to a mirror, their reflection reverses one word but not the other.

3. Two longer words that can be used for the same stunt are ELECTRONIC AUTOMATA.

5. ART, MUSIC, POETRY, AND NUMBERS

Reflection symmetry provides one of the oldest and simplest methods of creating a pleasing design. The child's inkblot pictures mentioned in the preceding chapter are examples of this. When a child is shown how to make one for the first time he usually squeals with delight when the sheet is unfolded and he sees the bilaterally symmetric pattern, especially if the picture is made with colored paints instead of dark ink. Why does a child think these pictures are "pretty"? The obvious answer is that he enjoys the sense of order or harmony that has been imposed on a random pattern. Is it because he sees so much bilateral symmetry in the world around him? No one really knows, though it seems reasonable to suppose that the bilateral symmetry of nature, so much a part of his experience, conditions him to respond with pleasure to this type of pattern. Bilateral symmetry is common in the art of primitive cultures and in the early history of painting. It was an essential aspect of the style of early Egyptian art. Medieval religious paintings were often designed with strong bilateral symmetry.

For modern tastes the composition of such a picture is dull because the symmetry is too obvious (although it seems to have had a temporary revival in the work of some of the "pop" artists, as well as in the geometrical paintings of recent abstractionists). Look around, however, and you will see endless examples of bilaterally symmetric shapes and patterns on man-made objects. I am not referring merely to objects that have such symmetry for reasons of convenience (doors, windows, chairs, and so on), but to patterns and shapes that are made symmetric primarily to make them pleasing. Vases, lamp bases, chandeliers, birdbaths, stained-glass windows, Christmas tree ornaments, earrings, lapel pins—the list is endless. Patterns on dresses, wallpaper, drapes, and tile floors are often a repetition of bilaterally symmetric figures. Trademarks and familiar emblems—the cross and *fleur de lis*, for example—usually have bilateral symmetry. As Hermann Weyl points out in his little book on *Symmetry* (Princeton University Press, 1952), resemblance to nature is sometimes completely sacrificed to obtain an exact duplication on each side of a vertical axis of symmetry. A striking instance is the double-headed eagle on the coat of arms of czarist Russia and the old Austro-Hungarian monarchy.

Japanese art, in marked contrast to Western and even Chinese art, strives to avoid symmetry. As Everett F. Bleiler points out in his introduction to Kakuzo Okakura, *The Book of Tea* (Dover, 1964), the Taoist and Zen notion of perfection, which so strongly influenced Japanese art, emphasizes dynamic growth, whereas symmetrical patterns suggest repetition and completeness. Japanese aversion to symmetry extends even into the ritual of serving tea. "In placing a vase on an incense burner on the tokonoma," Bleiler writes, "care should be taken not to put it in the exact center, lest it divide the space into equal halves."

In Western art, whenever bilateral symmetry is dominant, the patterns and shapes almost always have vertical axes of symmetry. We are so accustomed to vertical axes in the natural world that we would feel vaguely uncomfortable, without knowing exactly why, if certain wallpaper patterns with vertical-symmetry axes were rotated 90 degrees. There is, however, one common natural scene that *does* have a horizontal axis: the scene in which trees and other objects are reflected in a smooth lake or river. When we see a painting of such a scene, the feeling of discomfort is no longer present; the symmetry is pleasing (see Saul Steinberg's marvelous *New Yorker* cover, July 23, 1966). For similar reasons, lapel pins seldom have a single, horizontal axis of symmetry unless they resemble a fish or some other natural object usually seen in such a position.

The enormous preference that nature shows for vertical axes of symmetry is due, of course, to the simple fact that gravity is a force that operates straight up and down. As a consequence, things tend to spread out equally in all horizontal dimensions. Water spreads out to form lakes with horizontal surfaces. A lake is indifferent as to whether it spreads north or south, east or west, but is incapable of spreading up in the air. For this reason, if you take a photograph of a lake and reverse the negative to obtain a print in which right is left and left is right, it still looks like a perfectly ordinary lake. But if you turn the photograph upside down, the water is violating the law of gravity and you are seeing something that could not possibly occur in nature. A tree is, in a rough overall way, symmetric in the same way that a cone is symmetrical: it has an infinite number of vertical axes of symmetry, none that are horizontal. Again, gravity provides the obvious explanation. The tree grows upward against gravity. It has roots in the soil, leaves in the air. This distinguishes clearly its top from its base. Because it is rooted in the ground, and does not move from place to place like an animal, there is no front or back, no left or right. A mirror reflection of a tree, when we hold the mirror vertically, looks exactly like a tree.

In fact, it is difficult to tell if a photograph of *any* natural scene is reversed unless the picture happens to contain some bilaterally *asymmetric* man-made objects, such as a sign with printing on it or streets with cars driving on one side of the road. But if you reflect a photograph along a horizontal axis—which has the effect of turning it upside down—it is immediately apparent that something is amiss. In the *New Yorker* for May 5, 1962, page 189, appeared a cartoon showing a man, just out of bed in the morning, raising the shade of a bedroom window. The scenery, glimpsed through the window, is upside down. The cartoon is funny because this kind of reversal is so utterly preposterous; if the scene through the window were reversed from left to right it would look perfectly natural.

Now and then artists and cartoonists amuse themselves by drawing upside-down pictures—pictures that change to other pictures when inverted. The surprise experienced when such a picture is turned upside down is due to the fact that we never expect a picture to resemble anything at all when inverted. Left and right reversals are so commonplace that it is easy to imagine what a picture would look like if reversed, or "flopped," as the graphic artists like to say. It is almost impossible to study an upside-down picture and imagine what it looks like when turned around.

When I was in college I once found myself living in a furnished room hung with reproductions of pictures which I intensely disliked. To escape from their subject matter I turned all the pictures upside down. This obscured the subject matter, leaving only the colors and composition, which I found pleasing. Unfortunately, my landlady, who had bought the pictures and liked them, objected so strenuously that I had to turn the pictures right side up again. The point of the story is that inversion of a realistic picture, either by turning it around 180 degrees or by reflecting it in a mirror held horizontally above it (the two are not quite the same thing), certainly changes the aesthetic value of the picture. Is there any change in the aesthetic value of a picture when it is reflected left and right? One is tempted to answer no, but on second thought, there may be a subtle change arising out of the fact that most viewers, at least in Western nations, are accustomed to reading from left to right. Some critics of art have argued that a picture loses something of value when flopped.

There is some empirical evidence to support this. David B. Eisendrath, Jr., a New York photographer, once prepared a set of fifty scenic photographs so that each picture had two reproductions, one a mirror image of the other. The pairs were shown one at a time to various subjects who were asked to designate which one of each pair they liked best. Scenes that had an overall left–right symmetry were as often chosen in one form as the other, but if the scene showed a composition with strong asymmetry, there was about 75 percent agreement among subjects in the choice of one picture over its mirror twin. All these subjects read from left to right. When the same pictures were shown to subjects who read only Hebrew, which goes from right to left, there was a tendency to prefer the mirror reversals of those pictures that had been preferred by left-to-right readers.

These tests, and earlier work along similar lines by German psychologists (in particular, Heinrich Wölfflin and Theodora Haack), suggest that there may be some loss of aesthetic value when certain paintings are flopped. If so, the loss is certainly not great. You might try the simple experiment of going through a book that reproduces many paintings (preferably paintings you have not seen before), looking at each picture directly, then in a mirror to see if you detect any loss or increase in aesthetic pleasure.

When a set of pictures tells a story, as in the comic strip, the left-to-right order obviously has a strong influence on the way each picture is drawn. Action is usually from left to right, and the character who speaks first must be placed on the left to prevent the "balloons" from being read in the wrong order. On the Japanese makimono, a long strip on which pictures tell a story, the action goes in the reverse direction because the strip is unrolled from right to left.

A motion-picture film is easily reversed from left to right. You might watch such a film for some time before you realized that it had been flopped: perhaps you would catch sight of printing on an advertising sign or you would see two people shake with left hands. Statues are sometimes bilaterally symmetric (the general sitting on his horse), and bilateral symmetry in architecture is too familiar to call for any comments. In dancing, bilateral symmetry also plays a significant role. The Rockettes at Radio City Music Hall sometimes go through a routine in which left and right versions of almost every step alternate from start to finish.

Movies, in addition to being flopped, can also be reversed along the time dimension. They have a mad, nightmarish quality: people walking backward down the street, divers plunging out of the water onto diving boards, and so on. Would this also be true of the film of a ballet projected backward? Backward dancing might be rather pleasant, in a grotesque sort of way, particularly if synchronized with forward music. Could a skilled choreographer plan a palindromic ballet with bilateral time symmetry—that is, dancing that would appear almost the same if photographed on motion-picture film and projected backward?

One might suppose that mirror symmetry would play no role at all in music, but if we think of tones as sound patterns that flow along the single dimension of time, then a mirror reflection of a melody is obtained simply by playing the music backward. This is easily done with a tape recorder. In most cases the backward music is a meaningless jumble of sounds, not pleasant to hear. Piano music sounds strangely like organ music when reversed in time. (Can you guess why?) During the fifteenth century many composers constructed canons (songs like "Row, row, row your boat . . ." that consist of two or more melodies sung simultaneously) in which one melody is the other melody backward. Many of the greatest composers have used the reversal of melodies for various sorts of contrapuntal effects.

Music can also be turned upside down in the sense that high notes become low and low notes high. If you imagine an unreversed Alice, on the other side of the looking glass, sitting down at a piano and playing a familiar melody, this is the sort of music the reversed piano would produce. You can do the same thing yourself, if you have access to a player piano; simply turn a roll of music around to switch low and high notes, then play the roll forward from start to finish. In a joke canon, often falsely credited to Mozart, the second melody exhibits *both* types of reversal: that is, it is the same as the first melody turned upside down and read back to front. In this way only the one melody needs to be printed on a sheet of music. One person sings it with the sheet turned one way while the other person sings from the same sheet viewed upside down. A modern example of such a canon, constructed by Winthrop Parkhurst, appears in his book *The Anatomy of Music* (Knopf, 1930, page 137).

Poetry also may be thought of as a series of sounds ordered along the single dimension of time. There is no question but that many skillful poets have deliberately used reflection symmetry to achieve special sound effects. Robert Browning, for example, in his well-known, lovely lyric "Meeting at Night," employed a rhyme scheme of *abccba* so that the reflection of the sounds would give a feeling of the movements of sea waves:

> The grey sea and the long black land;
> And the yellow half-moon large and low;
> And the startled little waves that leap
> In fiery ringlets from their sleep,
> As I gain the cove with pushing prow,
> And quench its speed i' the slushy sand.

Then a mile of warm sea-scented beach;
Three fields to cross till a farm appears;
A tap at the pane, the quick sharp scratch
And blue spurt of a lighted match,
And a voice less loud, through its joys and fears,
Than the two hearts beating each to each!

If the forms of letters are disregarded and one thinks of a sentence as a series of symbols ordered along a straight line, then all sorts of amusing effects can be produced by mirror-reflecting the symbols. Palindromic words are words that are bilaterally symmetrical (spell the same in both directions): *radar, deified, rotator. Malayalam* is a language spoken in India. *Wassamassaw* is the palindromic name of a swamp in Berkeley County, South Carolina. A semordnilap ("palindromes" spelled backward) is a word that becomes a different word when reversed: *live, straw, desserts, redrawer.* When an entire sentence has bilateral symmetry it is called a palindrome. Thousands of remarkable palindromes have been composed. Two of the best:

A man, a plan, a canal—Panama!
Straw? No, too stupid a fad. I put soot on warts.

The first palindrome is sometimes attributed to the American humorist James Thurber, but this is not correct. It was composed by Leigh Mercer, a London word-puzzle expert who has composed many other excellent palindromes. The second palindrome is also Mr. Mercer's.[1]

A palindromic number is a number that remains the same when its digits are taken in reverse order. The last palindromic year, 1881, is also the same when turned upside down or held to a mirror. (1961 is invertible but not palindromic.) The next palindromic year is, of course, 1991. If you add any number whatever to the number obtained by reversing its digits, then do the same thing with the sum, and keep repeating, will you eventually reach a sum that is palindromic? Thus 89 plus 98 is 187, which is not a palindrome. 187 plus 781 is 968, still not palindromic. Keep going and eventually (in this case, after 24 additions) you reach the palindrome 8,813,200,023,188.

It has been conjectured that this procedure, applied to any integer, gives a palindrome after a finite number of steps. Charles W. Trigg, a California mathematician, doubts the truth of the conjecture. He has found 249 integers, each less than 10,000, which do not yield palindromes after 100 steps. The smallest of these integers is 196. Another California mathematician, Dewey C. Duncan, has shown that the procedure does not always produce a palindrome in the binary system; for example, when applied to the binary number 10110. For a proof that this number never produces a palindrome, see Problem 5 in Roland Sprague, *Recreation in Mathematics* (Blackie & Son, 1963).

In 1977 Heiko Harborth proved (in *Mathematics* Magazine, vol. 46, 1973, pages 96ff.) that the conjecture is false in all number notations with bases that are powers of 2. For all other bases the conjecture is still unresolved. In decimal notation, 196 was carried to 237,310 steps in 1975 by Harry J. Saal, at the Israel Scientific Center, without producing a palindrome.

Gustavus J. Simmons has written two papers about palindromic powers (*Journal of Recreational Mathematics*, April 1970 and January 1972). There are infinitely many palindromic squares, cubes, and fourth powers, but it is not known if there are palindromes that are powers higher than 4. The smallest palindromic square with a square root that is not also a palindrome is $676 = 26^2$. The largest known nonpalindromic number whose square is a palindrome is 3,069,306,930,693.

Palindromic cubes whose cube roots are not palindromes are so rare that only one is known: $10,662,526,601 = 2,201^3$. No palindrome is known that is an nth power, n greater than 3, whose nth root is not palindromic.

Palindromic primes have also received some attention. A prime is any integer, excluding 1, that is not evenly divisible by any other integer except itself and 1. A palindromic prime must begin and end with 1, 3, 7, or 9, and cannot have an even number of digits greater than 2 (otherwise it is a multiple of 11). Almost all interesting questions about palindromic primes are unanswered. It has not even been proved that there are an infinite number of such primes. A reversible prime is one that is not palindromic but is a prime when it is read backward. This set, too, is not known to be infinite or finite.

Norman Gridgeman has noticed that palindromic primes often

come in pairs that are identical except for the center digits, which differ by 1. For example, among the first 47 palindromic primes there are a dozen such pairs:

2	919	13831
3	929	13931
181	10501	15451
191	10601	15551
373	11311	16561
383	11411	16661
787	12721	30103
797	12821	30203

Are there an infinite number of such pairs? Gridgeman conjectures yes, but this also remains unproved.

Palindromic poems, in which the order of words is exactly the same in both directions, are written from time to time, and on rarer occasions, palindromic poems in which letters are the basic units. The following poem by Graham Reynolds is one of three palindromic poems that were printed in 1960 in *New Departures*, no. 2–3, an avant-garde magazine published at South Hinksey, near Oxford:

HYMN TO THE MOON

Luna, nul one,
Moon, nemo,
Drown word.
In mutual autumn
I go;
Feel fog rob all life;
Fill labor
Go, flee fog
In mutual autumn
I drown
Word: omen; no omen.
O, Luna, nul.

Frederic Brown has the distinction of having written an entire short-short-short story that is palindromic by words. He has given permission to reprint it in full, from his collection of outlandishly funny stories *Nightmares and Geezenstacks* (Bantam, 1961). I can think of no more appropriate closing for this chapter than Brown's

THE END

Professor Jones had been working on time theory for many years.

"And I have found the key equation," he told his daughter one day. "Time is a field. This machine I have made can manipulate, even reverse, that field."

Pushing a button as he spoke, he said, "This should make time run backward run time make should this," said he, spoke he as button a pushing.

"Field that, reverse even, manipulate can made have I machine this. Field a is time." Day one daughter his told he, "Equation key the found have I and."

Years many for theory time on working been had Jones Professor.

END THE

Notes

1. For more on palindromes, including palindromic poems and dramatic dialogues, see Howard W. Bergerson, *Palindromes and Anagrams* (Dover, 1973); C. C. Bombaugh, *Oddities and Curiosities of Words and Literature* (Dover, 1961); Dmitri Borgmann, *Language on Vacation* (Scribners, 1965) and *Beyond Language* (Scribners, 1967); my *Scientific American* columns for August 1970 and February 1977; and back issues of the quarterly *Word Ways*.

6. Galaxies, Suns, and Planets

The entire cosmos—the universe of space and time and everything it contains—seems to have, in a general, overall way, the symmetry of a sphere. We live on a small planet that revolves around the sun. The sun is one of a hundred billion stars that form our galaxy. The galaxy is of the spiral type, with long arms that trail outward from the center like the fiery arms of a monstrous pinwheel. Our solar system is in one of the arms, more than thirty thousand light-years (a light-year is the distance light travels in one year, or some six trillion miles) from the galaxy's center. The galaxy itself belongs to a cluster of galaxies. Beyond the cluster, at inconceivably vast distances, space is strewn with other galactic clusters. Astronomers have good reasons for thinking that these clusters are rushing away from each other so that the whole universe is expanding like a gigantic balloon that is being inflated.

According to one theory, known as the big bang theory, the amount of matter in the universe is finite. Billions of years ago it was all concentrated in one enormously dense lump. The lump exploded, and this explosion started the evolution of the universe. According to another theory, known as the steady state theory, the amount of matter in the universe is infinite. As it expands, new matter is constantly being created to prevent the universe from thinning out. In both theories the cosmos has an overall symmetry.

Consider, for a moment, the expanding universe of the big bang theory. Space may go on forever, but the material universe—the stars and other astronomical bodies that evolved after the primeval explosion—is finite and spherical in shape. Imagine an enormous plane slicing through the center of this universe. No matter what direction you slice it, the overall features of the universe on one side of the plane are like the overall features on the other side. In other words, the material universe has the symmetry of a sphere.

In the steady state theory the material universe stretches to infinity in all directions. There *is* no center. In this universe we can imagine an infinite plane that slices it in any direction, through any spot we please. Again, astronomers who accept this theory have no reason for thinking that the universe on one side of such a plane would be any different, in an overall way, from the other side. Such a universe would have the symmetry of an infinite, homogeneous 3-space. As in the big bang model, there is no indication that the universe has any type of large-scale left–right asymmetry.[1]

Are galaxies also symmetrical? Yes, even the spiral galaxies are symmetrical when we regard them as three-dimensional structures. It is true that a spiral *confined to a plane* is asymmetrical. There is no way to turn such a spiral over, to change it to its mirror image, without rotating it off the plane. But spiral galaxies are not plane figures. Viewed edgewise they have a lens shape, like two plates pressed together face to face. The plane that divides the "plates" is a plane of symmetry slicing the galaxy into mirror-image halves. As we saw in earlier chapters, this means that a spiral galaxy can be superposed on its mirror image. We simply turn one of the images around so both images spiral in the same direction. Of course we are still considering only the overall features of galactic structure. If we take into account the individual stars that make up the galaxy, their sizes and natures, and the patterns they form with each other, then the galaxy is not superposable on its mirror image.

There is a pseudo-sense in which a galaxy is not superposable on its reflection, even in an overall way; that is, when we take into consideration the north and south poles of its magnetic field. It is known that our galaxy has an extremely weak magnetic field. The exact structure of the field is not known, but it probably has a magnetic axis that coincides closely with its axis of rotation. If we take into account the labels we attach to the ends of such a magnetic axis, then the galaxy's "left" side is not the same as its "right." The galaxy is not superposable on its mirror image. If we turn one image around until its spiral arms mesh with the spiral arms of its enantiomorph, the magnetic north pole of one image will coincide with the south magnetic pole of the other. Actually, as we shall learn later, this is not a true asymmetry. It only seems so because of the way we label the ends of the magnetic axis. The magnetic field itself is symmetrical, but this cannot be made clear until, in chapter 19, we examine the nature of magnetism.

A similar pseudo-asymmetry holds with respect to stars such as our sun. When the shape of the sun is considered alone, it clearly has spherical symmetry. It is true that the sun rotates, but this does not prevent it from being superposable on its mirror twin. All we have to do is turn one image upside down, reversing the orientation of its axis of rotation, and the two images will coincide point for point, both rotating the same way. However, the sun is known to have a magnetic field. The magnetic axis, like the earth's magnetic axis, corresponds closely to the axis of rotation. If we label the ends "north" and "south," and do not reverse these labels on the sun's mirror image, the spinning sun and its reflection cannot be made to coincide. If the axes coincide, the spins will not; if the spins coincide, the axes will not. Curiously, and for reasons now inexplicable, the magnetic axis of the sun occasionally turns a complete somersault; the north pole becomes the south pole and vice versa. Since the sun does not alter its direction of spin, this magnetic flip-flop means that (in a sense) the sun changes handedness and turns into its own enantiomorph!

What about planets? Like the sun, they have spherical symmetry and are therefore superposable on their mirror images unless one takes into account irregular surface features or the pseudo-asymmetry of their magnetic field. The earth, of course, has such a field, with north and south magnetic poles that are not far from the north and south poles of the axis on which the planet turns. In addition to the pseudo-asymmetry introduced by its magnetic field, the earth also has a shape now known to be slightly (very slightly) like that of a pear. It used to be assumed that the earth was a perfect "oblate spheroid," a sphere slightly flattened at the poles, but accurate measurements in the last few years indicate that the flattening is a trifle greater at the south than at the north pole. If this variation is taken into account, then the spinning earth is like a spinning top in the sense that the shape of the upper part is not identical with the shape of the lower part, and (quite apart from the labels on its magnetic axis) it is not superposable on its reflection. If it were not *spinning*, there would be no asymmetry. Hold a top up to a mirror and you will see that its shape is the same as the shape of its mirror image. But as soon as you give it a spin, it acquires a "handedness." A top spinning clockwise (as you look down on it) has a mirror twin that spins counterclockwise. If you try to make the spins coincide by turning one image upside down, the two images will not mesh because the top of one image will be at the bottom of the other, and the top and bottom of a top are not the same.

On the surface of any rotating astronomical body, all sorts of interesting asymmetries develop that are "left-handed" in one hemisphere, "right-handed" in the other. For example, if you are in an airplane in the Northern Hemisphere, flying directly toward the North Pole, the pilot will have to correct for a marked tendency of the plane to deflect to the right as you face forward. If the plane is in the Southern Hemisphere, flying toward the South Pole, the deflection will be to the left as you face forward. This deflection is an instance of what physicists call the Coriolis effect, after Gaspard Gustave de Coriolis, an early nineteenth-century French engineer who was the first to study it thoroughly. The effect arises from the fact that an object, at different spots on the earth's surface, is carried through space at different velocities. If you stand at the equator, the rotating earth causes you to travel a circle of roughly 24,000 miles in twenty-four hours, giving you a velocity of 1,000 miles per hour. As you move toward one of the poles, you move to a spot where the circle you are traveling (as the earth spins) becomes smaller and smaller. Since you always complete the circle in the same twenty-four-hour period, your velocity through space must get less and less. At the pole, of course, the velocity becomes zero.

A similar variation in velocity depends on how far you are from the center of the earth. If you are on top of a high mountain, the circle you travel in twenty-four hours is larger than your circular path when you stand at the base of the mountain. As you walk down the mountain, your velocity of revolution grows less and less. It continues to decrease if you go down a mine shaft. The deeper you go, the slower you revolve. At the center of the earth your velocity would become zero.

It is not hard to see how these variations in speed would cause deflections of opposite handedness in the two hemispheres. Of course the deflection becomes significant only if objects travel vast distances at high speeds. In shooting a rifle at a target, the Coriolis deflection of the bullet would be too small to consider, but when intercontinental missiles are traveling north or south, the deflection becomes important and has to be taken into account if the firing is to be accurate. Imagine such a missile traveling through the Northern Hemisphere toward the North Pole. The farther north it travels, the smaller are the circles through which the earth is turning it. Because of the missile's inertia, it tends to *keep the original velocity with which it was moving eastward with the earth at the time of firing.* After it has traveled, say, 500 miles north, it is in a region where objects are being carried eastward by the earth at a much slower speed. But the missile retains its former eastward velocity. As a result it drifts east, or to the right, as it speeds toward the Pole. A little reflection will show that, if the missile is in the Southern Hemisphere, traveling toward the South Pole, the drift will be to the left as it moves forward. In both cases the drift is eastward, but if you plot the deflections on a globe you will see that each is a mirror image of the other.

In both hemispheres the Coriolis deflection is to the east as an object moves toward the Pole, to the west as an object moves toward the equator. It is not surprising that the Coriolis effect plays a significant role in the movements of atmospheric and oceanic currents. Some geologists think that rivers flowing north in the Northern Hemisphere and south in the Southern Hemisphere scour their eastern banks more than their western. There is no doubt that the Coriolis effect plays a role in the flow of rivers, but geologists disagree as to whether the effect is strong enough to be measurable by differences in erosion on the two banks. Some studies have been made of the banks of the Mississippi and other rivers that run north or south, but the results are debatable.

Also undecided is the question whether the Coriolis force is sufficiently strong to be detectable as an influence on water when it spirals down a drain. As everybody knows, when you let water out of a bathtub it forms a vortex around the drain. It is widely believed that the bathtub's vortex in one hemisphere has a handedness opposite to the vortex in the other. To see the reasoning behind this, consider a large, circular, flat-bottomed tub placed directly on top of the North Pole (Figure 18). The drain, at the tub's center, goes straight down into the earth. As water in the tub flows toward this opening, the Coriolis force deflects it eastward in the direction shown by the arrows, forming a counterclockwise vortex. Once the vortex starts, it reinforces itself and grows stronger; presumably, water flowing out of such a tub at the North Pole would tend to spiral down the drain in the counterclockwise fashion indicated.

At the South Pole the situation is reversed. True, the water still deflects eastward, but now such a deflection produces a clockwise vortex. This spiraling tendency of water down drains would be

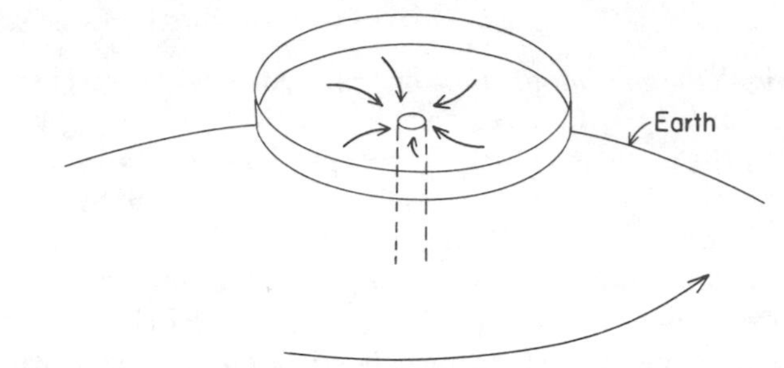

Figure 18. A bathtub vortex at the North Pole.

strongest at the poles, decreasing gradually as the tub moved toward the equator, where the tendency would vanish. In southern tubs the water would tend to "go widdershins" (the wrong way), as Scotch dialect has it. Tub water at the equator would behave like the proverbial donkey between two piles of hay; it would not know which way to go.

There is little doubt that if bathtubs not on the equator were large enough, and if the water were perfectly motionless before it started to drain, the earth's rotation *would* influence the handedness of the vortex. But bathtubs are small, and many other factors enter the picture. The strongest factor is the circulation the water acquires when the tub is filled. Water has an astonishingly long "memory"; circulatory movement will persist for hours, long after the water appears perfectly still. Such a circulation invariably starts the water spiraling down the drain in the same direction. Even if water is allowed to settle for several days before draining, the direction of the vortex may be influenced by slight irregularities in the surface of the tub, the sides of the drain, and so on.

Nevertheless, some tests seem to show that the Coriolis effect on draining bathtubs *can* be detected. In 1962 Ascher H. Shapiro, a physicist at M.I.T., made some experiments with a circular tub six feet in diameter. After letting the water settle for several days he found, when he pulled the plug, a consistent counterclockwise vortex which he attributed to the Coriolis effect. Merwin Sibulkin of General Dynamics, in "A Note on the Bathtub Vortex" (*Journal of Fluid Mechanics*, September 1962, pages 21–24), was unable to confirm this, possibly because he used too small a tub. With a circular tub one foot in diameter, and transparent sides so that the motion of dye in the water could be observed, Sibulkin found that the vortex always followed the circulation acquired by the water when the tub was filled, unless the water was allowed to settle many hours. No consistent counterclockwise rotation could be detected when completely quiescent water was drained. To his great surprise, however, he found that regardless of the handedness of the vortex, when the water had lowered to less than a half-inch in depth the handedness of the vortex mysteriously reversed. He suggests tentatively that the circulating water, as it drains, induces a reverse circulation in a layer of water on the bottom of the tub.

Shapiro's results were supported in 1965 (see *Scientific American*, November 1965, page 54) by five investigators at the University of Sydney. They obtained clockwise rotations after allowing water in a circular tub to settle eighteen hours or more before draining it at the center. "One can never prove," they wrote in *Nature*, "that it was not some small air current which persistently maintained a circulation that gave the results we observed, and that a quantitatively comparable, but oppositely directed, air current caused Shapiro's results. . . . Nevertheless, we have acquired confidence in the hypothesis that carefully performed experiments on liquid drainage from a tank will show clockwise rotation, if done in the Southern Hemisphere."

No one doubts that the Coriolis effect is responsible for the tendency of cyclones and tornadoes to spin counterclockwise in the northern half of the globe and to go widdershins in the other half. As for bathtub vortices, the question is still controversial, calling for bigger and better-controlled tubs before any final verdict can be rendered.

Notes

1. The steady state theory is no longer viable. It was discredited in 1965 by the discovery that the universe is permeated with a microwave radiation that can only be explained by assuming that it was produced by the primeval explosion that created the cosmos.

7. PLANTS AND ANIMALS

Among the billions of known galaxies scattered through space, each containing billions of stars, it seems reasonable to suppose that circling around many of these stars there must be planets, and that on some of these planets there must be life. "A sad spectacle!" exclaimed Thomas Carlyle, as he considered the possibility that the universe might contain planets by the millions. "If they be inhabited, what a scope for pain and folly; and if they be not inhabited, what a waste of space!"

At the moment, no one really knows whether life in any form is spread throughout the universe, confined to our own galaxy, or confined to our solar system. We do not even know if there is life on Venus or Mars, the two planets nearest the earth. The time is rapidly approaching, however, when some of these questions can be answered.

Assuming that forms of life have evolved on other planets, will these forms be wildly unlike anything that even science-fiction writers have imagined? Or will they possess certain features in common with life as we know it? It is all sheer speculation, of course, but with respect to questions of symmetry we can make some educated guesses. On the earth, life started out with spherical symmetry, then branched off in two major directions: the plant world with symmetry similar to that of a cone, and the animal world with bilateral symmetry. There are good reasons to think that evolution on any planet, if it occurs at all, would tend to follow a similar pattern.

Primitive one-celled life, floating in a sea and constantly tumbling about, would naturally assume a spherical form with planes of symmetry in all directions. But once a living form anchors itself to the bottom of a sea or to the land, a permanent up–down axis is created. The rooted end of any plant is obviously distinguishable from the upper end. There is nothing, however, in the sea or air to distinguish between the ends of a front–back axis or a left–right one. It is for this reason that plant forms, for the most part, have a rough, overall symmetry similar to that of a cone: no horizontal plane of symmetry, but an infinity of vertical planes. A tree, for example, obviously has a top and bottom, but one is hard put to distinguish the front from the back of a tree, or its right from its left. Most flower blossoms have, in a rough way, a conical type of symmetry. Fruits sometimes have spherical symmetry (if you ignore the spot where they attach to a branch): oranges, cantaloupes, coconuts, and so on. A cylindrical-type symmetry (an infinity of planes of symmetry passing through one axis, and one plane perpendicular to that axis and bisecting it) is exhibited by such fruits as grapes and watermelons. Familiar fruits with conical symmetry are the apple and pear. (Biologists use the term *radial symmetry* for symmetry of both cylindrical and conical types.) The banana furnishes an example of bilateral symmetry. Owing to its curvature and its pointed end, it is only possible to cut a banana into mirror-image halves by one plane of symmetry.

Are there examples of asymmetry (total absence of planes of symmetry) in the plant world? Yes, and the most striking examples are the plants that display helices in some part of their structure. As we learned in an earlier chapter, the helix cannot be superposed on its mirror image. It therefore has two distinct forms: the right-handed helix, which corresponds to a wood screw that turns clockwise as it enters wood; and the left-handed helix, which is the mirror image of a right-handed one. Helices abound in the plant world, not only in stalks, stems, and tendrils but also in the structure of myriads of seeds, flowers, cones, and leaves, as well as in the helical arrangement of leaves around a stalk.

It is in the climbing and twining plants that the helix can be seen in its most regular form. The majority of twining plants, as they coil upward around sticks, trees, or other plants, coil in right-handed helices, but there are thousands of varieties that coil the opposite way. Some species have both left- and right-handed varieties, but usually a species has its own handedness, which never varies. The honeysuckle, for example, always twines in a left-handed helix. The bindweed family (of which the morning glory is a well-known species) always twines in a right-handed helix. When two plants of the same handedness twine around each other, the result is a fairly orderly production of interwound helices, all of the same type; when plants of opposite handedness coil around each other, they produce a hopeless tangle. The mixed-up violent left–right embrace of the bindweed and honeysuckle, for example, has long fascinated English poets. "The blue bindweed," wrote Ben Jonson in 1617, in his *Vision of Delight*, "doth itself enfold with honeysuckle." Shakespeare, in act 4, scene 1 of *A Midsummer-Night's Dream*, has Queen Titania describe her intended embrace of Bottom the Weaver (whose top has been transformed by Puck into the head of an ass) by saying: "Sleep thou, and I will wind thee in my arms. . . . So doth the woodbine the sweet honeysuckle gently entwist. . . ."[1]

More recently, a charming song about the love of the bindweed for the honeysuckle has been written by Michael Flanders, a left-handed London poet and entertainer, and set to music by his friend Donald Swann. On a visit to the Natural History Museum in Kensington, Flanders had been struck by an exhibit dealing with the left- and right-handed habits of climbing plants. The result was his song *Misalliance*. (You can hear it sung by Flanders and Swann on the Angel recording of their engaging two-man revue, *At the Drop of a Hat*.) With Flanders's permission, I quote the lyrics in full:

MISALLIANCE

The fragrant Honeysuckle spirals clockwise to the sun[2]
And many other creepers do the same.
But some climb counterclockwise, the Bindweed does, for one,
Or *Convolvulus*, to give her proper name.

Rooted on either side a door, one of each species grew,
And raced towards the window-ledge above.
Each corkscrewed to the lintel in the only way it knew,
Where they stopped, touched tendrils, smiled, and fell in love.

Said the right-handed Honeysuckle
To the left-handed Bindweed:
"Oh let us get married
If our parents don't mind, we'd
Be loving and inseparable,
Inextricably entwined, we'd
Live happily ever after,"
Said the Honeysuckle to the Bindweed.

To the Honeysuckle's parents it came as a shock.
"The Bindweeds," they cried, "are inferior stock.
They're uncultivated, of breeding bereft.
We twine to the right, and they twine to the left!"

Said the counterclockwise Bindweed
To the clockwise Honeysuckle:
"We'd better start saving,
Many a mickle maks a muckle,[3]
Then run away for a honeymoon
And hope that our luck'll
Take a turn for the better,"
Said the Bindweed to the Honeysuckle.

A bee who was passing remarked to them then:
"I've said it before, and I'll say it again,
Consider your offshoots, if offshoots there be.
They'll never receive any blessing from me."

Poor little sucker, how will it learn
When it is climbing, which way to turn.
Right–left–what a disgrace!
Or it may go straight up and fall flat on its face!

Said the right-hand thread Honeysuckle
To the left-hand thread Bindweed:
"It seems that against us all fate has combined.
Oh my darling, oh my darling,
Oh my darling Columbine,
Thou art lost and gone forever,
We shall never intertwine."

Together they found them the very next day.
They had pulled up their roots and just shrivelled away,
Deprived of that freedom for which we must fight,
To veer to the left, or to veer to the right!

In addition to coiling around things in a helix of a certain handedness, twining plants also have stems that twist in the same way they coil. Sometimes two or more stems of the same plant will twine together in ropelike fashion. The bignonia, for example, tends to form triple strands that twist to the right; the honeysuckle tends to form double strands that twist to the left. At times the trunks of beeches, chestnuts, and other trees exhibit a violent twisting of the bark into helical patterns, though the twist may be either to the right or left regardless of the species.

Sessile animals (animals attached to something and unable to move about on their own power), such as the sea anemones, usually have a conical type of radial symmetry like that of most plants. Slow, weakly moving animals, such as the echinoderms (starfishes, sea cucumbers, and other species) and jellyfish, likewise have conical symmetry. These animals float about in the sea or lie on the bottom where food and danger approach them with equal probability from all sides. However, as soon as a species evolved strong powers of locomotion it was inevitable that features would develop that would distinguish the animal's front from its back. In the sea, for example, the ability to move about rapidly in search of food gave an animal a great competitive advantage over sessile and slow-moving forms. A mouth is obviously more efficient on the front end of a fish than on its back end; the fish can swim directly toward food and gobble it up before some other animal gets it. This single feature alone, the mouth, is sufficient to distinguish the front end from the back (or, as biologists like to say, the *cephalic* from the *caudal* part) of a fish. Other features, such as eyes, also are clearly more efficient at the front end, near the mouth, than at the back. A fish wants to see where it is going, not where it has been. In short, the mere fact of swimming through water brought about a situation in which it was inevitable that forces of evolution would devise features that would distinguish one end of a sea animal from the other.

At the same time that locomotion was leading to distinctions between front and back, the force of gravity was causing similar differences between an animal's top and bottom, or, to use the biologist's terms again, the dorsal and ventral. (When an animal such as man stands upright, then of course his dorsal and ventral sides correspond to back and front, and his cephalic and caudal ends become top and bottom, but in this section we are confining our attention to sea life.) What about right and left? A moment's reflection and you will realize that there is nothing in the sea's watery environment to make a distinction between right and left significant. A swimming fish encounters a marked difference between forward and backward because one is the direction it goes, the other is the direction it comes from. The fish also encounters a marked difference between up and down. If it swims up, it reaches the surface of the sea. If it swims down, it reaches the ocean floor. But what difference does it encounter if it turns left or right? None. If it turns left, it finds the sea, and the things in it, exactly like the sea that it finds if it turns right. There are no forces, like the force of gravity, that operate horizontally in one direction only. It is for these reasons that various features—fins, eyes, and so on—tended to develop equally on left and right sides. Had there been a great advantage for a swimming fish to see only to the right and not the left, no doubt fish would have developed only a single eye on the right. But there is no such advantage. It is easy to understand why a single plane of symmetry remained, dividing fish bilaterally into mirror-image right and left sides.

When the reptiles crawled out on the land and evolved into birds and mammals, there was nothing in their new environment to call for a change in bilateral symmetry. Up and down now became an even stronger influence on an animal's structure, because appendages were needed for locomotion across the ground. Feet are of little value attached to the back of an animal and sticking up in the air! Of course the difference between front and back continued to be important.[4] As for left and right, the situation on land or in the air remained as symmetrical as in the sea. An animal in the jungle or a bird in the sky finds its environment on the left pretty much like its environment on the right. It is easy to understand why the bodies of land and air animals preserved the bilateral symmetry they had previously acquired in the sea. H. S. M. Coxeter, in his beautiful book *Introduction to Geometry* (Wiley, 1961), reminds us that it may have been this bilateral symmetry that William Blake described in those familiar lines:

> Tyger! Tyger! burning bright
>   In the forests of the night,
> What immortal hand or eye
>   Dare frame thy fearful symmetry?

In view of the overall symmetry of the earth and the forces acting upon it, it is hard to conceive of circumstances in the future that could alter this fundamental type of symmetry in the bodies of animals. The slightest loss of bilateral symmetry, such as the loss of a right eye, would have immediate negative value for the survival of any animal. An enemy could sneak up unobserved on the right.

We are now in a position to understand why, if there are animals on another planet, capable of moving through its seas, through its atmosphere, or over its land, it is likely that they, too, will have bilateral symmetry. On another planet, as on earth, the same factors would operate to produce such symmetry. Gravity would provide a fundamental difference between up and down. Locomotion would create a fundamental difference between front and back. The lack of any fundamental asymmetry in the environment would allow the left–right symmetry of bodies to remain unaltered.

Can we go further than this? Can we expect more detailed similarities of extraterrestrial life with life as we know it? Yes, we can. In the strange seas of another planet, regardless of their chemical composition, it is hard to imagine a simpler form of locomotion for evolution to exploit than the motion achieved by waving tails and fins. That evolution would find this type of propulsion is supported by the fact that even on the earth it has developed independently. Fish developed tail-and-fin propulsion. Then fish evolved into amphibious forms that crawled out on the land and became reptiles. The reptiles developed into mammals. But when some of the mammals returned to the sea—those that eventually became whales and seals, for example—their legs evolved back into flippers and their tail into a finlike instrument for propelling and steering.

Similarly, it is hard to imagine a simpler mode of flying through the air than by means of wings. Again, even on earth there has been independent, parallel development of wings. The reptiles evolved wings and became airborne. So did the insects. Some mammals, like the flying squirrel, developed wings for gliding. The bat, another mammal, developed excellent wings. A species of fish, leaping out of the water to escape capture, developed rudimentary gliding wings. Even man, when he builds an airplane, builds it with "wings" on a pattern that resembles a bird in flight.

On land, is there a simpler method by which an animal can move about other than by means of jointed appendages? The legs of a dog are not much different in mechanical working from the legs of a housefly, although they had a completely independent evolution. Of course the wheel also is a simple machine for moving along the ground, but there are good reasons why it would be difficult for a wheel to evolve. For one thing, it needs to be supported by an axle; either the wheel must be detached from the axle and free to turn on it, or the axle itself must turn and therefore be detached from the body. Then there is the huge problem of finding a way for the body to rotate a wheel. The difficulties are great, though I suppose not insurmountable. L. Frank Baum in *Ozma of Oz* invented a race of men called the Wheelers who had four legs like a dog, each terminating in a small wheel instead of a foot. In *The Scarecrow of Oz* he invented the Ork, a bird with a propeller on the end of its tail. If on some planet nature found a way of inventing the wheel, we might find there animals resembling bicycles and cars, fish resembling motorboats, and birds resembling airplanes, although the prospects seem most unlikely.[5]

Sensory organs such as eyes, ears, and noses also have about them a kind of inevitability if life evolves any type of advanced intelligent activity. Electromagnetic waves are ideal for giving a brain an accurate "map" of the outside world. Shock waves transmitted by molecules provide additional valuable clues to the environment and are picked up by ears. The spread of actual molecules from a substance is detected by noses.[6] Since light, sound, and molecules certainly exist on other planets, it seems likely that evolution would invent senses to exploit these phenomena as a means of achieving greater control over the circumstances of life. Here on earth, for instance, the eye has had no fewer than three quite independent, parallel developments: the eyes of vertebrate animals, the eyes of insects, and the eyes of various mollusks. The octopus, for example, has a remarkably good eye—in fact, in some respects it is superior to our own. It has eyelids, cornea, iris, lens, retina—as does the human eye—yet it evolved entirely independently of the evolution of the vertebrate eye! It is hard to find a more astonishing instance of how evolution, operating along two disconnected lines of development, managed to invent two complicated instruments that have essentially the same function and structure.

There are good reasons for eyes and other sensory organs to form a kind of face. In the first place, there is an advantage in having eyes, ears, and nose close to the mouth, where they are useful in the search for food. There is an equally great advantage in having them close to the brain. It takes time for a nerve impulse to get from the organs to the brain; the quicker it gets there, the quicker an animal can react in catching food or avoiding danger. Even the brain itself, needed to evaluate and interpret sensory data, accomplishes its thinking by electrical networks; a kind of miniature electrical computer of great complexity. Nerve filaments that carry electrical charges may be essential for the brains of advanced living creatures.

If life on another planet reaches the intelligence level of man on earth, it seems probable that it would have at least a few humanoid features. There are obvious advantages in having fingers at the ends of arms. For protection, the valuable brain would need to be heavily encased and as far from the ground as possible, where it would be best shielded from the shocks of moving about. Sensory organs, close to the brain and in front, would create something like a face. "Senator" Clarke Crandall, a Chicago entertainer, had a funny routine about the advantages of having sensory organs at other spots on the body. An eye on the tip of a finger, for example, would make it possible to see a parade by holding up a hand and looking over the heads of everybody. Ears under the armpits would be kept warm in cold weather. A mouth on top of the head would allow a man to put a sandwich under his hat and eat it on the way to work. It is easy to see why evolution has avoided such arrangements. An eye on the finger would be too vulnerable to injury, too far from the brain. Armpit ears would not be very efficient for hearing unless you kept your arms perpetually raised. A mouth on the head would expose the brain to injury, make it difficult to see what one was eating, and so on.

Of course so many chance factors are involved and environments of planets are so varied that one would not expect to find on another planet any form of life that was a close replica of any species on earth. No one expects to find an elephant or a giraffe on Mars. On the other hand, alien life may not be so wildly different from earth forms as one is tempted to think. The BEMs of science fiction (BEM is an acronym formed by the initials of Bug-Eyed Monster), unlike any earthly animal but nevertheless recognizable as animals, may prove to be not far from the truth after all. It is hard, in fact, to imagine how extraterrestrial creatures could differ from earth creatures to any greater degree than earth creatures differ from each other. The octopus, the platypus, the hornbill, the ostrich, the snake—if one had never seen or heard of these animals, their structure would seem as bizarre and improbable as that of any animal we are likely to find on Mars. We have a fine specimen of a miniature BEM in the anableps, a small, bluish Central American carp that has four eyes! Well, not really. The huge eyes, like monstrous bubbles, are divided into upper and lower halves by an opaque band. Each eye has a single lens, but there are upper and lower corneas and irises. The little fish (it is about eight inches long) swims with the opaque band exactly at water level, so that its two upper "eyes" can see above water while its two lower "eyes" see under water. In the next chapter we will learn something about the asymmetric sex life of this curious creature.

Animals as weird as the anableps, no doubt much weirder, likely roam the seas, land, and skies of alien planets, but they are not likely to be so unearthly that we do not recognize them as animals. The chief basis for this recognition, more fundamental than any other aspect of their forms, is likely to be the bilateral symmetry of their bodies.

Notes

1. In Shakespeare's day the bindweed was sometimes called the woodbine. Later *woodbine* became used exclusively as another term for honeysuckle, a fact that has confused dozens of easily confused Shakespeare commentators. Some of them have even reduced the passage to silliness by supposing that the beautiful Queen Titania, "sometime of the night," was speaking of herself and Bottom as entwined like honeysuckle with honeysuckle. Awareness of the opposite handedness of bindweed and honeysuckle heightens, of course, the meaning of Titania's passionate metaphor.

2. In this book I have adopted the convention of calling a helix right-handed if it corresponds to the helical thread of a common wood screw. Flanders adopts the opposite convention of calling such a helix left-handed because, when you look at it from either end, you see it spiraling *toward you* in an anticlockwise direction. This confusion of terminology runs through all the literature on climbing plants.

3. A Scottish phrase meaning "many a little makes a lot."

4. There are some amusing exceptions to this in ancient mythology and modern fantasy. The *amphisbaena* (in Greek it means "go both ways") was a fabled Greek snake with a head at each end. It crawled both ways. Here is how Pope described it in his *Dunciad:*

> Thus Amphisbaena (I have read)
>   At either end assails;
> None knows which leads, or which is led,
>   For both Heads are but Tails.

In recent fantasy for children there is Duo, the two-headed dog in L. Frank Baum's *John Dough and the Cherub*, and the Pushmi-Pullyu of Hugh Lofting's Dr. Dolittle books. Both animals had a head at each end.

5. Although no known animal uses a wheel for propelling itself along the ground or through the air, there are bacteria that move through liquids by actually rotating flagella like propellers. (See "How Bacteria Swim," by Howard C. Berg, in *Scientific American*, August 1975, page 36ff.) There may be rotary devices inside cells for unwinding twisted strands of DNA. (See *Scientific American*, February 1967, page 37.) Some one-celled animals propel themselves through water by rotating their entire body. Nor must we overlook the dung beetle, the sacred scarab of Egypt, that transports little balls of dung by rolling them across the ground.

6. It is not impossible that there may be advanced cultures of intelligent nonterrestrials in which smell and taste not only are the dominant senses but also provide the primary means of communication between individuals. Only in recent years have biologists discovered how much information, in terrestrial animal species, is transmitted efficiently by a direct transfer of substances now called pheromones. See Edward O. Wilson's nose-opening report on "Pheromones" in *Scientific American*, May 1963.

8. ASYMMETRY IN ANIMALS

Just as left–right asymmetry turns up here and there in the radially symmetrical world of plants, so also does it turn up in the bilaterally symmetric world of animals. An entire volume could easily be devoted to these asymmetries. We have space only to discuss a few of the most interesting examples.

As in the plant world, asymmetry is automatically introduced whenever a single helix forms part of the structure of an animal. Of course, when a helix on one side of an animal's body is balanced by a helix of opposite handedness on the other side , bilateral symmetry is preserved. This applies to pairs of tusks that have helical twists (for example, the tusks of extinct mammoths), and to the large magnificent horns of rams, goats, antelopes, and other animals. Many large bones in the chest, legs, and other parts of animals (including man) have helical twists, but those on the left side have their mirror counterparts on the right. Insect antennae sometimes coil in pairs of enantiomorphic helices. The wings of birds, bats, and insects also have slight helical twists of opposite handedness on opposite sides of the body.

When a single helix is prominent in the structure of an animal, then an obvious asymmetry exists. Many types of bacteria and the spermatozoa of all higher animals have such helical structures, but the most striking examples are provided by the shells of snails and other molluscs. Not all spiral shells are asymmetrical. The chambered nautilus, for instance, coils on one plane and therefore can be bisected, like a spiral nebula, by a plane of symmetry. But there are thousands of beautiful molluscan shells, such as those shown in Figure 19, that are obviously either right- or left-handed conical helices. As in the case of twining plants, most shells of this type are right-handed, but both types of handedness are common. Some species are always right-handed, some always left-handed. Some are right-handed in one locality, left-handed in another. Each species has occasional "sports" that go the wrong way; they are rare and much prized by shell collectors. Thousands of different species of fossil shells, with right- or left-handed helices, have been classified by the paleontologists.

An odd type of helical fossil known as the devil's corkscrew is found in great abundance in parts of Nebraska and Wyoming. These huge

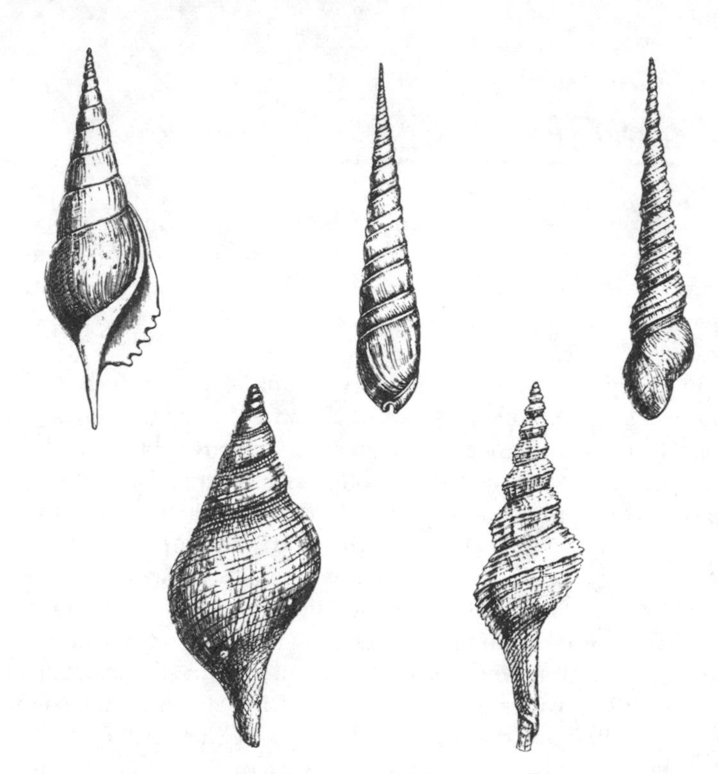

Figure 19. Right-handed mollusc shells.

quartz spirals, six feet or more in height, are sometimes right-handed, sometimes left-handed. For decades, geologists argued with each other about what they were, the chief division being between those who thought they were fossils of long-extinct twining plants and those who thought they were casts of helical burrows made by ancestors of modern beavers. The beaver theory finally won out after remains of small beavers were found in some of them. Similar spiral fossils, of similar origin, are found in parts of Europe.

A remarkable instance of helical flight is furnished by the hundreds of thousands of Mexican free-tailed bats that sleep in the limestone caverns of Carlsbad, New Mexico. Joseph Wood Krutch in his book *The Desert Year* (Sloan, 1952) gives a vivid description of how these bats, when they swarm out of a cave, invariably gyrate in a counter-clockwise spiral. Mr. Krutch wonders just how the bats managed to agree on which type of helix to adopt. "Their convention is certainly a 'socially useful one,' " he writes. "Without it, a bat would find leaving the cave almost as dangerous as driving to work in a car."

Is it possible that Coriolis forces have something to do with this: that bats tend to emerge from Northern Hemisphere caves in left-handed helices, from Southern Hemisphere caves in right-handed helices? Krutch checked with a number of leading bat authorities but was unable to find any significant information bearing on the question. A Coriolis influence seems highly unlikely; nevertheless, the handedness of helical paths taken by emerging bats remains an interesting area that seems not to have been explored by the naturalists. "Perhaps someday someone will turn a discarded wind tunnel on end," says Mr. Krutch, "and put a few hundred bats at the bottom of it. . . . The bats have got into my belfry . . . I can already see my application to one of the foundations. Proposed Project: 'A Study of the Coriolis Effect in Relation to Bat Flight.' "

Turning to animal asymmetries other than helical, one of the most absurd is the huge left or right pincer of the fiddler crab (Figure 20). The crab makes a kind of fiddling motion with this pincer, which gives the crab its name. Among birds, a pleasant example of asymmetry is provided by the crossbill, a small red bird in the finch family. The bird's upper beak crosses over the lower beak like the blades of a pair of scissors, and, like scissors, can cross in either of two mirror-image

Figure 20. The left-handed fiddler crab.

ways. The species dominant in the United States has the upper bill crossing to the bird's left; the dominant European species has a bill that "goes the other way." The bill is used for prying open evergreen cones in much the same way that a plierlike instrument is used by housewives for prying off the lids of cans. Once the cone is open, the bird sticks in its tongue and extracts the seed. A colorful ancient legend has it that the bird took pity on the crucified Jesus and with its bill tried to pull the nails from the cross. This vain effort of mercy twisted the bird's beak and stained its plumage with blood. The only bird whose entire beak twists to one side is the wry-billed plover of New Zealand. It uses the beak to turn over stones when looking for food, and since the beak bends to the right, the bird looks for food mainly on the right.

The female birds of all genera, with few exceptions, exhibit a curious left–right asymmetry with respect to their ovaries and oviducts. In young birds, both the left and the right ovaries and their ducts are equal in size; as the bird matures, the organs on the right degenerate and become useless. Only the left oviduct, which greatly enlarges during the egg-laying season, remains functional.

In the fish world the outstanding instance of asymmetry is supplied by the flatfish, a large family which includes the soles and flounders. The young of these fish are bilaterally symmetrical with an eye on each side. They paddle about near the surface of the sea, but as they grow older, one eye slowly migrates around over the top of the head until both eyes are on the same side, like the eyes in the profile of a face painted by Picasso. The poor fish then sinks to the bottom of the sea, where it lies in the mud or sand, on its eyeless side, with its two eyes projecting upward. The eyes turn independently; the fish can look forward with one, backward with the other. The blind underside of the fish is whitish, but the upper side is colored and speckled to imitate the bottom of the sea. Some species even have the power of altering their color to conform to wherever they are lying and thus better escape detection by enemies. There are hundreds of different species of flatfish, most of them with eyes invariably on the right side, others that always have their eyes on the left. The halibut, for example, is a dextral, or right-eyed, flounder; the turbot is a sinistral, or left-eyed, flounder. There are dextral soles found only in European waters, sinistral soles found only in tropical and semitropical waters.

In every species an occasional "sport" will differ in handedness from his cousins. There is an interesting discussion of flatfish in chapter 7 of Charles Darwin's *Origin of Species.* (Darwin replies effectively to a critic of evolution who maintains that there is no conceivable way by which the peculiar migration of the flatfish eye could take place as a result of natural selection.)

The anableps, the little "four-eyed" fish mentioned at the end of the preceding chapter, has an asymmetric sex life which is absolutely unique among vertebrates. Its young are born alive, which means that a male must fertilize the female's eggs inside her body. But the female has an opening only on the left or right side, and the male organ is also only on the left or right. In other words, each individual fish is sexually either sinistral or dextral, making it impossible for two fish of the same handedness to mate. Fortunately, the handedness of both males and females is mixed fifty-fifty; if both sexes had the same handedness, the species would soon be in serious trouble. Here we have an amusing analogy in the fish world to the mating of the bindweed and honeysuckle.

The tusks of animals (tusks are simply teeth that have been enlarged to serve some special purpose), such as those of the elephant and walrus, are seldom exactly the same size. Usually a species tends to be either right- or left-tusked, in the sense that one tusk is a bit larger than the other and more often used. In Africa, the right tusk of elephants is sometimes called the "servant" tusk, because the elephant prefers to use it for digging.

The narwhal, a species of small whale that flourishes in north polar seas, exhibits the most extreme example of asymmetric tusk development. Both sexes of the narwhal have only two teeth; they lie side by side, on either side of the plane of symmetry, within the creature's upper jaw. In the female narwhal both teeth stay permanently inside the jawbone. The right tooth of the male remains similarly concealed throughout life, but the left tooth grows straight forward into an ivory tusk which is longer than half the whale's length! If the whale has a length of twelve feet from tail to snout, this ridiculous tooth will be seven or eight feet long and as straight as a spear. It is, in fact, the longest tooth in the world (see Figure 21).

Around the tusk are helical grooves and ridges, which always spiral forward in a counterclockwise direction. On rare occasions, *both* teeth

Figure 21. Skull of the narwhal, seen from below.

of a male narwhal may grow into tusks. When this happens, one might expect that, like the horns on rams and goats, one tusk would have right-handed grooves, the other left-handed ones. But no, both tusks invariably coil in the same left-handed way! This has long puzzled zoologists. One theory, advanced by Sir D'Arcy Thompson in his famous book *On Growth and Form* (an abridged edition was published by Cambridge University Press in 1961), rests on the fact that the narwhal swims forward with a slight screw motion to the right. The inertia of the tooth would tend to keep it in place while the body twisted, thus imparting to the tooth a torque that would cause it to rotate slowly counterclockwise as it grew forward.

"The horn does not twist round in perfect synchronism with the animal," Thompson writes, "but the animal (so to speak) goes slowly, slowly, little by little, round its own horn! The play of motion, the lag, between head and horn is slight indeed; but it is repeated with every stroke of the tail. It is felt just at the growing root, the permanent pulp, of the tooth; and it puts a strain, or exercises a torque, at the very seat, and during the very process of calcification." Thompson's theory has been criticized, but so far no biologist seems to have found a better one.

The narwhal is sometimes called a sea unicorn because of its single "horn." In fact, during the fifteenth and sixteenth centuries the creature's tusks were sold throughout Europe, mainly by Scandinavian traders, as horns of actual unicorns. Powder made from such a horn was widely believed to have all sorts of miraculous prophylactic properties. The racket was finally exposed by a Dutch zoologist in the early seventeenth century.

Exactly what purpose the giant tooth serves remains to this day a mystery. There is no evidence that it is ever used for stabbing enemies, as early zoologists thought, or for punching through ice to make breathing holes. During the mating season male narwhals sometimes cross horns with each other, like a pair of fencers, so it may be that the tooth's only purpose is to serve as part of a sex ritual.

There are thousands of other striking instances of animal asymmetries: the way wings overlap on crickets, grasshoppers, cockroaches, and other insects; the asymmetric ears of certain owls that help them locate sound origins; the akita dog in Japan with a tail that curls one way on males, the other way on females; the tendency of dolphins to swim counterclockwise around tanks; the asymmetric sex organ of the male bedbug; a fungus called *laboulbeniales* that grows only on the back left leg of a certain beetle. A. C. Neville, a British zoologist, has collected many more examples in his splendid little book *Animal Asymmetry* (Edward Arnold, 1976).

The human body, like the bodies of most animals, has an overall bilateral symmetry coupled with minor deviations from symmetry. The topic is sufficiently curious and complicated to call for a separate chapter.

9. THE HUMAN BODY

The unclothed human figure displays an almost flawless bilateral symmetry. Certainly part of the aesthetic appeal of a well-proportioned nude, in the flesh or in a work of art, derives from the mirror-reflection identity of the body's right and left sides. (The female figure shows no asymmetry. Male symmetry is broken only by the curious fact that the left testicle usually hangs lower than the right.) Of course, any individual body may have minor deviations from symmetry: one shoulder higher than the other, a slightly twisted spine, a scar or birthmark on one side of the body, and so on; but such deviations, for the most part, are as likely to be on one side as on the other.

Bilateral symmetry persists in the body's interior, especially in the muscles and skeleton, but is broken in many spots by the grossly asymmetric placing of various organs. The heart, stomach, and pancreas are shifted leftward; the liver and appendix are on the right. The right lung is larger than the left. Twists and turns of the intestines are completely asymmetric. The human umbilical cord, a magnificent triple-helix of two veins and one artery, invariably coils counterclockwise (see Figure 22).

Figure 22. The human umbilical cord.

Ordinary twins, who develop from the simultaneous fertilization of two separate egg cells, can have asymmetric features which go one way on one twin, the other way on the other, but this happens no more often than would be expected by chance. It is widely believed that identical twins—twins who develop from a single egg that divides soon after fertilization—have a marked tendency toward mirror imaging. Unfortunately, the statistics are fuzzy and many authorities believe that identical twins are no more likely to mirror one another than are ordinary siblings.

In the case of Siamese twins—identical twins joined to each other as a result of late, incomplete division of the egg—there is no doubt about the matter. They are exact enantiomorphs in almost every detail. If one is right-handed, the other is left-handed.[1] If the hair on the crown of the head of one whorls clockwise, the hair of the other whorls counterclockwise. Ear differences, tooth irregularities, and so on appear on both twins in mirror-image forms. Palm prints and fingerprints of one twin's right hand will be closer to the prints of the other twin's left hand than to his own left hand. The same is true of his other hand.

Even more startling, one Siamese twin will have a "transposed viscera": his internal organs will be reversed—heart on the right, liver on the left. This transposition of the viscera, or *inverse situs*, as it is sometimes called, is always found in one of a pair of Siamese twins, but it may also occur in singly born individuals. It is much rarer than dextrocardia, in which only the heart and major blood vessels are reversed. When it occurs outside of twinning it is usually associated with other physical abnormalities, such as harelip, cleft palate, extra fingers and toes. Readers who wish to know more about Siamese twins and their astonishing mirror imagery should consult chapter 5 of Horatio Hackett Newman's *Multiple Human Births* (Doubleday, 1940), a fascinating, popularly written book by a famous University of Chicago biologist and expert on twinning.

It is worth noting that Lewis Carroll, in *Through the Looking-Glass,* intended Tweedledee and Tweedledum, that well-known pair of identical twins, to be taken as mirror images of each other. When the Tweedle brothers offer to shake hands with Alice, one extends his right hand, the other his left. If you study Tenniel's illustrations carefully, especially the picture in which the twins face each other arrayed for battle, you will see that he has drawn them as if they were enantiomorphs.

In the behavior and habits of human beings there are many examples of marked asymmetry, the most obvious deriving from the fact that most people are right-handed. The right hand is controlled by the left side of the brain and the left hand by the right side of the brain, so right-handedness is actually a phenomenon of left-brainedness. At one time it was thought that babies are born with no genetic tendency to favor either hand; that the handedness of a child is solely the result of parental training. This view is strongly expressed by Plato.

"In the use of the hand we are, as it were, maimed by the folly of our nurses and mothers," Plato writes in Book Seven of his *Laws,* "for although our several limbs are by nature balanced, we create a difference in them by bad habit." Favoring one hand over the other is of little consequence, the Greek philosopher goes on to say, in such tasks as playing the lyre, which must be held in one hand and plucked with the other. But in such sports as boxing and wrestling, especially in hand-to-hand battle combat, it is essential that a man learn to use both hands with equal skill. For this reason, he argues, children should be trained to use both hands equally for all tasks.

Today we know that Plato was badly mistaken. As Aristotle correctly pointed out, our arms are *not* balanced by nature. An inherited tendency for most people to favor the right hand is universal throughout the human race and for as far back as history provides reliable evidence. Cultural anthropologists have yet to find a society or even a local tribe in which left-handedness is the rule. The Eskimos, the American Indians, the Maoris, the Africans—all are right-handed. The ancient Egyptians, Greeks, and Romans were right-handed. Of course, if you go far enough back in history the evidence for right-handedness becomes scanty and indirect. It has to be deduced from such clues as the shapes of tools and weapons, and pictures showing men at work and in battle. In drawing a face in profile, a right-handed person finds it easier to draw the face facing left, a fact that also serves as a clue to the handedness of prehistoric men. Anthropologists who have investigated the handedness of primitive man do not agree among themselves, so no firm conclusions can be drawn, but there is no disagreement about the right-handedness of all societies since the beginning of recorded history.

The very words for left and right, in most languages, testify to a universal right-handed bias. Our word "right" suggests that it is right to use the right hand. It may be that "left" had its origin in the fact that the left hand is so little used that it is "left out" of most tasks. When a compliment is given with malicious intent, we call it a left-handed compliment. *Sinister*, suggesting something disastrous or evil, is from the Latin word for left; *dexterous* or *dextrous*, meaning adroit and skillful, is from the Latin word for right. The French word for left is *gauche*, which also means crooked or awkward; the French word for right is *droit*, which also means just, honest, and straight. Our word *adroit* is based on the French word. In German the word for left is *link* and the word *linkisch* means awkward. The German *recht* for right means just and true, as it does in English. The Italian left hand is called the *stanca*, which means the fatigued, or the *manca*, which means the defective. Spaniards speak of the left hand as *zurdo*, and the Spanish phrase *a zurdas* means the wrong way. Ojo the Unlucky, the boy protagonist of L. Frank Baum's *Patchwork Girl of Oz* is left-handed, as was Baum himself. "Many of our greatest men are that way," the Tin Woodman tells Ojo. "To be left-handed is usually to be two-handed; the right-handed people are usually one-handed."

Christianity has strongly reinforced the West's association of left with evil. It is difficult to find a praiseworthy reference to the left hand anywhere in the Bible. When Joseph's father blesses Joseph's two sons, he confuses left with right and accidentally (much to Joseph's distress) puts his right hand on the second-born instead of on the first. Jesus spoke of separating the sheep from the goats at his Second Coming, the sheep going to the right, goats to the left. The devil was supposed to be left-handed. Saints were said to have refused their mother's left breast when they were nursing. Painters of the Last Judgment showed God pointing to heaven with his right hand, to hell with his left.

The prejudice has been just as strong in the East, particularly in Japan. Left-handed pupils in rural areas were often beaten by teachers, and left-handed girls had to pretend to be right-handed to find a husband. In 1968 Soichi Hakozaki, a Tokyo psychiatrist, tried to combat the prejudice (which he believes causes much needless anxiety) with a book called *Warnings Against Rightist Culture. Time* (January 7, 1974) reported Hakozaki as gratified by the sales of his book but fearful that it was a "long uphill battle." One encouraging sign was a Japanese hit record titled "My Boy Friend Is a Lefty."

No one knows why the entire human race has this built-in inherited tendency to favor the right hand. Monkeys and apes, our closest cousins among the primates, are ambidextrous. Some vertebrates show a left–right preference in certain respects—pointer dogs that point with one paw, parrots that perch on one leg, and so on—but all this is too remote from the human race to be relevant. At some time in the geologic past, when primates began the great transition to human types, something started them off on this asymmetric habit. It has been pointed out that in fighting an enemy, primitive man may have found it an advantage to carry a knife or spear in his right hand, where it would have a minimum distance to travel before piercing the heart of his adversary. In addition, his more vulnerable left side would need the protection of a shield. The shield would naturally be carried by the left hand, leaving the right free to grip the weapon. A mutation favoring the right hand might, in the light of these factors, have a slight survival value. Lee Salk has a theory that newborn babies need to hear the mother's heartbeat, so primeval mothers held their babies in their left hand, freeing the right to perform various tasks. Other theories have been put forth to account for right-handedness, but there is little evidence to back up any of them. Most anthropologists consider it a mystery not yet satisfactorily accounted for.

What percentage of the general population today is left-handed? It seems a straightforward, simple question; actually it is vague almost to the point of being meaningless. One could write an elementary book on statistics by surveying critically the enormous, confusing literature that has been published in the last few decades on this question. In the first place, the incidence of left-handedness may vary from time to time, area to area. In the second place, it is not easy to define exactly what is meant by "left-handed." It is true that most people are right-handed, but among those who are not, some are strongly left-handed, some are weakly left-handed, some are ambidextrous in the sense that they do almost everything equally well with both hands, some are ambidextrous in the sense that they are equally clumsy with both hands. Some perform certain skilled tasks with their right hand, other skilled tasks with their left. It is not uncommon to find a person who writes with one hand, but eats and does everything else with his other hand. Or vice versa. Finally, it is extremely difficult to identify a person who may have been born with a bias toward left-handedness but never showed the bias because he was trained from an early age to use his right hand.

In view of such difficulties, it is not surprising to find the experts in wild disagreement over the incidence of left-handedness. In fact, estimates vary all the way from 1 percent to more than 30 percent! One of the earliest records relevant to this question is in the Old Testament: chapter 20 of Judges, verses 15 and 16. The passage is not clear, but it seems to say that out of 26,000 men in an army, 700 left-handers were chosen who could "sling stones at a hair breadth, and not miss." The passage is interesting because it suggests that the left-handers were unusually skillful, and because it gives the percentage of left-handers at about 2.7. Today, most studies show a much higher incidence. Many authorities estimate that about 25 percent are born left-handed but that the environmental pressures of a right-handed world cut the sinistral minority to a much lower fraction.

Newsweek magazine, in its October 1, 1962 issue, included a questionnaire designed to find out how many readers habitually read the magazine from back to front and to see if such a practice had any correlation with left-handedness. An analysis of 5,800 replies appeared in the February 25, 1963 issue: 56.1 percent read from front to back, 43.9 percent from back to front. If this survey is to be trusted, it is surprising that so many Westerners have adopted the habit of reading a news magazine from back to front, the way magazines are designed to be read in the Orient. There was no significant correlation of this habit with left-handedness. Among back-to-front readers, 13 percent reported themselves left-handed, 85.1 percent right-handed, 1.9 percent ambidextrous. Among front-to-back readers, 12.4 percent said they were left-handed, 84.8 percent right-handed, 2.8 percent ambidextrous. Thus, among the total number of *Newsweek* readers who responded, about one-eighth were left-handed.

There are many indications that the incidence of left-handedness has risen in the United States during the last few decades. Most authorities think that this is not because more people are being born left-handed but because parents have become more permissive in allowing left-handers to remain left-handed. Thirty or forty years ago parents were told by psychologists that all sorts of nervous disorders, especially stammering, might result if a left-handed child were taught to use his right hand for eating and writing. Not only would the changeover put a child in a state of emotional stress and rebellion, but (some authorities maintained) his brain would become confused as to which side was dominant, a confusion that would implicate the brain's speech centers.

Today the consensus among experts is that matters of handedness play an extremely minor role, if any, in stuttering and similar nervous disorders. Wendell Johnson, professor of speech pathology and psychology at the University of Iowa's famed speech clinic, has written a fine book called *Stuttering and What You Can Do About It.* In this book Johnson summarizes the strong evidence that has led psychologists to abandon the once widely held theory that there is a connection between stammering and handedness. Careful and impressive studies have disclosed no link of any sort.

Dr. Johnson himself stammered as a child, and there is a funny-sad section of his book in which he describes the succession of vain efforts he made to find a cure. He tried faith healing, speaking with pebbles in his mouth, adjustments by a chiropractor, three months in a stammering school where he swung dumbbells while reciting such lines as, "Have more backbone and less wishbone." He ended up at the University of Iowa, where a new program on stuttering was under way. The psychiatrists in charge were convinced that stuttering was related to handedness. There was no evidence whatever that Johnson was anything but strongly right-handed, but so well-anchored was the current theory that for ten years he tried to turn himself into a left-hander, totally without success! When the new data began to come out in the thirties, showing no correlation of stammering with handedness, Johnson himself could hardly believe it.

News of the new point of view is slowly filtering down to today's parents. Most child psychologists advise today that if a child is not strongly left-handed, no harm is done by gently and kindly inducing him, if possible, to switch to his right hand for eating and writing. If the child is strongly left-handed, however, it is best to allow him to remain so, not because compelling him to switch will make him a stammerer, but because it will emotionally upset him and probably be unsuccessful anyway. What effect such emotional stress may have on a left-hander, in the hands of overpersistent parents, is a topic still being debated.

Most right-handers are right-footed, in the sense that they usually kick a ball with their right foot, but in other respects tend to be left-footed. In situations where the left foot is most often used, the tendency may be correlated with right-handedness. In mounting a horse, for example, making the left leg do the work allows the right arm to do most of the assisting. If a spade is held with the right hand on the end, where it can do the strongest pushing, it is more convenient to use the left foot for forcing the spade into the ground. Right-handed boys usually mount a bicycle from the left. I would guess that most right-handers leap forward by making their left leg do the muscular work, but I have seen no statistics to support this. When a person is lost in the woods, he tends to circle either clockwise or counterclockwise, under the impression he is walking in a straight line. There have been attempts to correlate this with footedness, but the results are inconclusive.

Most right-handers are also right-eyed, in the sense that their right eye dominates in seeing. A simple eyedness test is to focus your eyes on a distant spot, then raise your finger until an image of the fingertip (necessarily out of focus) coincides with the spot. Since there are two images of the out-of-focus finger, one for each eye, you will tend to use the image of your dominant eye. By winking first one eye and then the other you can tell which image you picked. Most people use their dominant eye for looking through a microscope or telescope. Whether the dominant eye is most often used for winking at someone is a question about which no research seems yet to have been undertaken.

For testing the dominant eye of a small child, ophthalmologists use all sorts of costly instruments, but you can make an excellent one in just a few minutes. Simply roll a sheet of paper into the form of a small megaphone. Tape it so it doesn't unroll. Ask the child to peek at you through the *large* end of the megaphone. The eye that *you* see through the megaphone is his dominant eye.

Psychologists have also turned up evidence that a right-handed person tends to be right-eared and right-jawed. That is, his right ear dominates his hearing, and he tends to use the right side of his mouth more than the left for chewing. There also is some evidence that when a right-hander carries something heavy on his shoulder he uses his left shoulder. On the other hand, there seems to be no evidence of any correlation between handedness and the way a person applauds, clasps his hands, folds his arms, or crosses his legs. In each case the maneuver has a mirror-image form. Every person habitually chooses only one of the two forms (try clasping hands or folding arms the "wrong" way and note how queer it feels), but the choice seems to have no strong tie-in with handedness.

The right and left cerebral hemispheres of the human brain look superficially alike, but on closer inspection they have marked external differences that are probably genetically determined. What about the brain's interior—that mysterious minicomputer that processes information in ways nobody yet understands? It is now known that the interior mechanisms of the two sides of the brain function in startlingly different ways.

The left and right hemispheres are joined by a large bundle of nerves called the corpus callosum. When this cable is cut, persons who suffer extreme forms of epilepsy are enormously improved. In the 1950s Roger W. Sperry and his colleagues began a series of ingenious experiments with patients who had undergone this "split-brain" operation. Since then many other clever researchers have entered the field. There is little doubt that the left hemisphere is primarily concerned with language in all forms: speaking, writing, hearing, and reading words. It is also the side of the brain that seems to process data sequentially, like a digital computer. It is the side that thinks in logical, analytical, mathematical ways. The right hemisphere is more concerned with gestalts, with recognizing holistic patterns such as melodies, works of art, the structure of a human face. It appears to work in parallel fashion. It is the side of the brain that makes intuitive leaps. It may even be the side that processes emotions.

Split-brain research is only in its infancy, but already a great deal of nonsense is being written about it by psychologists, especially parapsychologists, who oversimplify and go far beyond established fact to spin fanciful theories. In an April-fool hoax column for *Scientific American* (April 1975) I described a "psychic motor" that rotates either clockwise or counterclockwise when a hand is held near it. Several parapsychologists sent serious letters suggesting that the direction of rotation depends on which side of the brain is dominating the psi force. Harold Puthoff and Russell Targ, in their book on clairvoyance (*Mind-Reach,* 1977), conjecture (page 102) that psychic abilities are a function of the right hemisphere. This notion, which had earlier been put forth by Robert Ornstein in his *Psychology of Consciousness,* is rapidly becoming popular among believers in psi phenomena. Skeptics, they like to say, are blinded to the reality of psi by their left-brain dominance.

Since each side of the brain controls the opposite side of the body, some writers on the paranormal have speculated that psychics, being right-brained, tend to be left-handed. Is not Uri Geller, the greatest of modern "psychics," a lefty? But correlations, if any, between handedness and the divisions of labor in the brain remain obscure. There is, however, no doubt that left-handers do not have an easy time of it in the right-handed cultural world, as we shall see in the next chapter.

Notes

1. The opposite handedness of Siamese twins is involved, in a minor way, in a confusing series of left–right clues in Ellery Queen's *Siamese Twin Mystery* (Frederick A. Stokes, 1933).

10. THE SINISTRAL MINORITY

Unless you yourself belong to the "sinistral minority," you probably do not fully realize the extent to which a strong left-hander finds it awkward to function at his best in a right-handed society. In most sports that use asymmetric equipment, fortunately, the lefty can buy items especially designed for him: fishing reels, baseball gloves, golf clubs, bowling balls, and so on. In 1968 a shop called Anything Left-handed, Ltd., carrying every available device for the southpaw, opened in London's West End.[1] Some banks issue special checkbooks for left-handers. A left-handed dentist can buy dental equipment that allows him to stand on the patient's left. All this is to the good, but in many ways the lefty is still unavoidably penalized. He has to write on a sheet of paper from left to right. If he sits at a lunch counter, especially in Manhattan, where the seats are closer together than in other large cities, his left arm is in perpetual conflict with the right arm of his neighbor on the left. Scissors, wall pencil sharpeners, wall can openers, salad forks, egg beaters, comptometers, and dozens of other familiar devices are designed for right-handers and are therefore awkward for the lefty to operate. The spiral notebook has its helix positioned to give the southpaw maximum discomfort.

A left-hander is constantly annoyed in all sorts of other little ways by the world's right-handed bias. When he takes a subway he finds the slot for the token on the right side of the turnstile. When he enters a phone booth he finds the door designed to be opened by a right hand. Inside the booth is a pay phone with a receiver to be held by the left hand, freeing the right hand for depositing coins, dialing, and taking notes. Has it ever occurred to you (if you are right-handed) that all wristwatches are made for right-handers? Try winding a wristwatch placed on your right wrist and you will see how awkwardly the stem is placed for left-handed winding. Instruction manuals are invariably written for right-handers. A left-handed girl who wants to learn how to knit, or a left-handed boy who wants to take up card conjuring, has to translate every left into right, every right into left. For more on the tribulations of a left-hander in a right-handed world, see James T. deKay's amusing *Left-Handed Book* (Evans, 1966).

How important these inconveniences are in conditioning the personality of the lefty is another matter on which there is little agreement among experts. During the period in which it was fashionable to equate stammering with left-handedness, it was also fashionable to equate all sorts of neurotic adult behavior with left-handedness. An outstanding instance is provided by Florence Becker Lennon's admirable biography *The Life of Lewis Carroll.* Although there is no documentary evidence that Carroll, a right-hander, was born left-handed, Miss Lennon deduces that this must have been the case on the basis of his lifelong habit of stuttering and the fact that his nonsense humor relied heavily on a technique of logical inversion. "If Charles was reversed," she writes, "he took his revenge by doing a little reversing himself. . . . The function of the left-handed person is thus to hold the mirror, and though such a nature may develop either stubbornness or perversity as its keynote, Charles leaned to the perverse rather than to the stubborn adaptation."

The suggestion that left-handers tend to be stubborn or perverse has been widely believed in the past, but few contemporary psychologists hold with it. The belief reached its height in the views of nineteenth-century criminologists, especially Cesare Lombroso, the Italian psychiatrist and crime expert. Lombroso was convinced that a higher proportion of left-handers were to be found in prisons than in the general population, and he wrote extensively in defense of his view that left-handedness was one of the degeneracy signs of the born criminal.

Today Lombroso's views are universally considered pseudoscience, but there may be a germ of truth in the correlation of crime and left-handedness if it is interpreted as an environmental phenomenon. In the nineteenth century, before parents developed permissiveness about left-handedness in children, there may have been many bitter conflicts between strongly left-handed youngsters and parents who tried to beat them into the use of their right hand. It is easy to understand how such conflicts might have led to difficulties that would predispose a person toward crime. A few modern criminologists who favor this view report that many left-handed criminals speak in interviews of having been severely punished by parents for using their left hand, but one suspects that prisoners are inclined to go along with any suggestion that transfers blame for their behavior. Statistics on this are extremely shaky. One must conclude that the correlation between crime and left-handedness, if there is such a correlation, is a topic not yet adequately investigated.

In some professions there are slight but unmistakable advantages in being left-handed; for that reason one would expect to find in those professions a higher incidence of lefties. Consider baseball. A left-handed pitcher is said to be confusing to a right-handed batter, and a left-handed batter (for example, Babe Ruth) is said to be similarly confusing to a right-handed pitcher. A left-handed batter, because he stands on the right side of home plate, actually has a shorter distance to run to first base. A left-handed first baseman has an advantage in having his catching hand on the right; it enables him to stand a bit closer to first and still cover his portion of the infield. In many other sports, such as tennis, boxing, and especially fencing, the southpaw has an advantage over right-handed opponents.

John Scarne, the gambling expert, discloses (in his book *Scarne's Complete Guide to Gambling*, Simon and Schuster, 1961) that there is a high incidence of left-handedness among professional blackjack dealers. The reason is rather subtle. A common method by which a dealer can cheat the customer has to do with taking a secret peek" at the top card's index. He can then either deal the top card or hold it back by surreptitiously dealing the second card. Owing to the fact that the indices of a playing card are asymmetrically placed (top left and lower right corners), a left-handed dealer, holding the deck in his right hand, can execute a "peek" that is superior to the types of peeks available to him when the deck is held in the other hand.

A factor that may tend to throw a higher proportion of lefties into professions such as baseball and card hustling is the simple fact that in certain professions there is no *disadvantage* in being left-handed. A strongly left-handed child finds writing and drawing with the left hand extremely awkward. It is hard to see what has been written and hard to keep the fingers from smudging the page, especially if ink is used. Only a few written languages, such as Chinese, Japanese, Hebrew, and Arabic, go right to left across the page. A left-handed child might well tend to dislike all activities that involve a great deal of writing or drawing. He would prefer instead such fields as music and sports, where left-handedness either does not matter or may even be an asset. Unfortunately, accurate statistics about the incidence of left-handedness in various professions are hard to come by. What proportion of artists and architects are left-handed? What proportion of violin players? Violins are designed to be bowed by the right hand, but the left does just as much, if not more, work. Do professional jugglers tend to be ambidextrous? Professional magicians? It would be interesting to have detailed statistics on these questions.

In every society a wide variety of social customs reflects the dominant right-handed bias: shaking hands, pledging allegiance, saluting, taking an oath of office, making religious gestures, and so on. A lefty must, of course, adapt to all of them. Walking on the right side of sidewalks and stairways, with such corollaries as the counterclockwise rotation of revolving doors, seems to be more of a convention than a practice tied up with right-handedness. The convention of driving cars on the right side of the road is now almost universal throughout the world. The British Isles and India are the main exceptions, but it is becoming increasingly difficult for these countries to resist the pressure for change. Imported cars have steering wheels on the left (for right-of-the-road driving), and tourists caught in a conflict of habit patterns are constantly causing accidents. Sweden, where the tourist accident rate was especially high, was the last nation on the continent to switch from left to right driving. When the big changeover took place—on Sunday, September 3, 1967 (after four years of careful and costly preparation)—a leftist newspaper in Malmoe celebrated the occasion by printing a "rightist" edition that read from right to left.

At the end of Vladimir Nabokov's novel *Lolita*, Humbert Humbert drives alone down a highway. Since he has disregarded all laws of humanity he decides he might as well disregard traffic laws, so he switches to the road's left side. The feeling is good. "It was a pleasant diaphragmal melting, with elements of diffused tactility, all this enhanced by the thought that nothing could be nearer to the elimination of basic physical laws than deliberately driving on the wrong side of the road. In a way, it was a very spiritual itch. Gently, dreamily, not exceeding twenty miles an hour, I drove on that queer mirror side."

Perhaps it is the right-side driving rule that is behind the convention, at least in the United States, of counterclockwise travel around the field in so many sports: car racing, horse racing, bicycle racing, dog racing, roller derbies, skating rinks, track events, and so on. Baseball players dash counterclockwise around the bases. Carousels and carnival rides usually go counterclockwise. As correspondent Scot Morris put it, "It seems that everything goes counterclockwise except clocks." But even clock hands go counterclockwise, someone has said, if you consider their motion from the clock's point of view.

Is it possible that right-side-of-the-road driving and walking accustoms us to seeing things go by from left to right? In a car, the nearest side of the road, the side easiest to see, goes past us from left to right. Standing on a sidewalk, or looking out a window, we see the traffic (if it is a two-way street) go from left to right. Would this make us slightly uncomfortable if we were to see horses go around a track the opposite way, or to see scenery spin past us the wrong way on a merry-go-round? It occurs to me that it is easier to hop on a moving merry-go-round by grabbing a pole with the right hand than the left, a fact that might explain its direction of spin in a right-handed world. Do carousels indeed go counterclockwise in all cultures?[2] Do horses race counterclockwise in all cultures, including those in which cars drive on the left side? Which way did Roman chariots race around the Circus Maximus? It would be instructive to have some accurate worldwide statistics on such questions.

Ancient inscriptions sometimes read from left to right, sometimes from right to left. The ancient Greeks had a curious form of writing called *boustrophedon* ("as the ox turns") in which the lines alternated left to right and right to left; one's eyes followed a continuous, snakelike path from top to bottom. Today the asymmetric convention of reading and writing from left to right is universal throughout the Western world. It is difficult for most right-handers to write in mirror form from right to left, and there is evidence that the ability to do this comes most easily to a person who is strongly left-handed. Leonardo da Vinci, a famous left-hander, could write mirror script as easily as, if not better than, he could write the usual way. In fact, he kept all his notebooks in mirror-image script, partly because this made it harder for snoopers to read them.

As an experiment, take a pencil in hand and see if you can write your signature from right to left so that it can be read correctly in a mirror. Is it any easier to do when you try it with your other hand? If you have access to a blackboard, try writing your name simultaneously with both hands from left to right with your right hand, from right to left with your left hand. You will probably find that your left hand does the job with greater facility this way than when only the left hand is used. Another trick for writing your name in reverse form is to put a sheet of paper on your forehead and with your usual hand write your signature from the left side of your head to the right. On the paper, of course, the signature will go the wrong way.

For another interesting experiment in left–right habits, pile some books in front of a dresser mirror. The stack must be high enough to screen your gaze from a sheet of paper placed on the dresser top between the books and the mirror. On the paper you have previously drawn a simple geometrical figure such as a spiral or a five-pointed star. By looking in the mirror, try to trace the figure by moving the point of a pencil along its lines. You will find it surprisingly frustrating. The reason of course is that you have learned to coordinate your hands with the image your eyes and brain give you of the world. When that image is reflected it is not easy for your brain to go against its learned reflexes and send your hands the correct signals.

Clothing in all societies tends to follow the bilateral symmetry of the human body, although many absurd deviations are decreed by custom. In some cases they may have been conditioned by right-handedness, For example, the lapel buttonhole on a man's jacket is always on the left side, perhaps because a right-hander finds it easier to put a flower on the left lapel than on the right. For a similar reason, one guesses, women favor the left side for lapel pins. Rings are usually worn on the left hand, where they do not interfere with handshakes and kitchen chores. The small change pocket of a man's jacket is inside the right pocket for easy access by the right hand.

A curious left–right custom in the Western world has to do with the way in which the coats of men and women overlap. On men's clothes the overlap is from left to right, with buttons on the right side, buttonholes on the left. The reverse is true of women's coats. Double-breasted trench coats are sometimes made with buttons and buttonholes on both sides so the coat can be worn by either a man or a woman. Of course a man would button it one way, a woman the other way. In 1963 this was the basis for the primary clue in a Perry Mason TV murder mystery. The murderer had rifled the victim's pockets, then rebuttoned the dead man's trench coat. Bette Davis (the substitute lawyer for an ill Perry Mason) cracked the case by noticing that the trench coat was buttoned the way a woman buttons it. This convinced her that the chief suspect, a man, was innocent.

Exercise 7: *What is the flaw in Bette Davis's reasoning?*

This concludes our survey of left–right symmetry in large-scale physical and biological structures. In the next chapter we take our first plunge into microscopic and submicroscopic levels, where left and right asymmetry is as omnipresent as it is in the macroscopic world.

Notes

1. Shops for lefties can be found in many large cities. The Aristera Organization (9 Rice's Lane, Westport, Connecticut 06880) issues catalogs for a large variety of southpaw items including playing cards, mustache cups, soup ladles, rulers, guitars, wristwatches, fishing reels, irons with properly positioned cords, and notebooks with the spiral binding on the right. The Left Hand, Inc., in Manhattan at 140 West 22nd Street, New York, N. Y. 10011, carries similar items.

2. In Ray Bradbury's fantasy novel *Something Wicked This Way Comes*, there is a carousel that carries riders back in time when the carousel is rotating the wrong way.

11. CRYSTALS

Our explorations of symmetry and asymmetry in nature began with the largest of all natural objects, the universe itself; we have been gradually moving down the scale of size to smaller and smaller structures. Two of the previous chapters were concerned with the overall symmetry of plants and animals. At this point we turn our attention toward still smaller structures, the various subunits that make up all material substances, living and nonliving.

Before going any further it will be good to have a clear understanding of just what these subunits are. Starting with the smallest first, then moving upward, the ladder goes like this:

1. The elementary particle. This is the smallest unit of structure known. The most important elementary particles, because they are the units of ordinary matter, are the proton, neutron, and electron.

2. The atom. This is the smallest structural unit into which matter can be divided and still have the properties of matter. At the center of every atom is the nucleus; it must contain at least one proton but usually contains a mixture of protons and neutrons. Around the nucleus, arranged in "shells," are the electrons. The simplest atom, hydrogen, has a nucleus of one proton, around which a single electron moves. The most complicated atom found in nature (more complicated ones can be created in the laboratory) is the uranium atom. It has ninety-two electrons.

Protons have a positive electric charge, electrons a negative charge. Neutrons, as the name suggests, are neutral; they have no charge. Ordinarily, the number of protons in an atom is the same as the number of electrons, so the charges balance each other and the atom itself is neutral. If an atom loses an electron from its outer shell, it becomes positively charged. If it gains an extra electron in its outer shell, it becomes negatively charged. Charged atoms are called ions.

3. The molecule. This is the smallest structural unit into which a chemical substance can be divided and still have the properties of that substance. When the substance consists entirely of one type of atom, the substance is called an element. In the case of certain rare gases, such as helium and neon, the molecule is simply one atom, but usually a molecule contains two or more. The molecule of hydrogen, for example, is made up of two atoms of hydrogen; a molecule of oxygen consists of two atoms of oxygen.

When atoms of different kinds are united in a molecule, the substance is called a compound. Ordinary water is a compound. Its molecule contains two atoms of hydrogen united by chemical bonds with one atom of oxygen. The atoms in the molecule of a compound can vary in number from two or three to the tens of thousands of atoms that form a single molecule of a complex protein.

4. The crystal unit. When the molecules of any solid substance are arranged in a fixed geometrical pattern, the substance is called a crystal. The pattern keeps repeating itself throughout the substance, like a two-dimensional pattern on wallpaper or a linoleum floor. Just as you can look at patterned wallpaper, point to a unit of design, and say, "That is a basic unit that keeps repeating," so you can examine the three-dimensional pattern of a crystal and find a basic arrangement of molecules that keeps repeating.

This is as high as the ladder of the structure of matter goes. Of course we can speak of still larger units, such as minerals and rocks, but no new type of mathematical pattern enters the picture. A mineral is merely an element or compound, in solid form, found in nature and not the result of some living process. But if a mineral exhibits a geometrical structure, it is a crystalline structure deriving from the arrangement of its molecules. Rocks are simply mixtures of one or more different minerals. Of course rocks do sometimes show a kind of pattern, such as the horizontal layers of sedimentary rock, but the pattern is of such low order that symmetry questions, of the type with which we are concerned, do not enter the picture.

Now that this quick survey of the four pattern levels is out of the way, we can return to our exploration of mirror symmetry. We will begin at the top of the ladder, with the structure of crystals, then in subsequent chapters work our way down the ladder into the subatomic jungle of the elementary particles.

Only solids have crystalline structures. The molecules of a gas are so far apart from each other that they are free to move haphazardly; it is impossible to find an orderly geometric pattern in their arrangement. The molecules of a liquid are closer together but still sufficiently free in movement to prevent the formation of fixed patterns. Solids, on the other hand, have molecules that pack together tightly to create a rigid, stable structure. (Actually, the atoms in a solid continue to oscillate, but the electromagnetic forces grip them so tightly that their oscillations are about fixed positions. For our purposes we can assume that the atoms are not in motion.) In almost every case, such an arrangement of molecules is patterned. This orderly pattern is the crystalline structure of the solid.

Consider water. In both gaseous form (steam) and liquid form its molecules are in a state of disorder, but when water freezes into a solid, the molecules group themselves into a pattern. The beautiful snow crystal, with hexagonal symmetry like a pattern in a kaleidoscope, takes its shape directly from the underlying crystalline pattern of ice molecules. Because of this underlying pattern, ordinary ice, from the ice cubes in your refrigerator to the mammoth icebergs in the Arctic, has a crystalline structure.

Almost all solid substances are crystalline. Glass is one of the outstanding exceptions. It is formed by cooling certain liquids so rapidly that the molecules freeze into a tightly packed position before they have a chance to arrange themselves in any orderly way. "Solid or not," write Alan Holden and Phyllis Singer in their excellent paperback *Crystals and Crystal Growing* (Anchor, 1960), "a glass is not crystalline. The cut-glass punch bowl, which is 'crystal' to the shopkeeper, is not crystal to the physicist. 'Crystal gazers,' who used to look into the future through spheres polished out of large single crystals of quartz, often look today through spheres of glass, because they are cheaper. It would be interesting to know whether the future seems as clear through a disorderly material as through an orderly one."

Noncrystalline solids are called amorphous; some chemists speak of them as liquid solids because, like liquids, they lack a crystalline structure. Charcoal, tars, and certain plastics are other familiar examples. Such substances share with liquids a tendency to "flow," though the rate of flow may be extremely slow. Even glass itself will flow out of shape if left alone for a few hundred years.

The underlying geometrical pattern of any crystalline substance is called the lattice of that substance. Sometimes the lattice is an arrangement of atoms, sometimes an arrangement of molecules. Carbon dioxide, for example, is found in nature as a gas; it is part of the atmosphere. When its temperature is lowered sufficiently it freezes and becomes what is known as dry ice. (It is called dry because it never melts, like ordinary ice, into a liquid; it just turns directly back into a gas.) In dry ice the molecules of carbon dioxide group themselves into the cubical lattice shown in Figure 23. Cubical lattices, similar in structure to the steel girders of an office building, are the simplest types of lattices. The molecules at the face of each cube give this particular lattice the name of *face-centered cubic.*

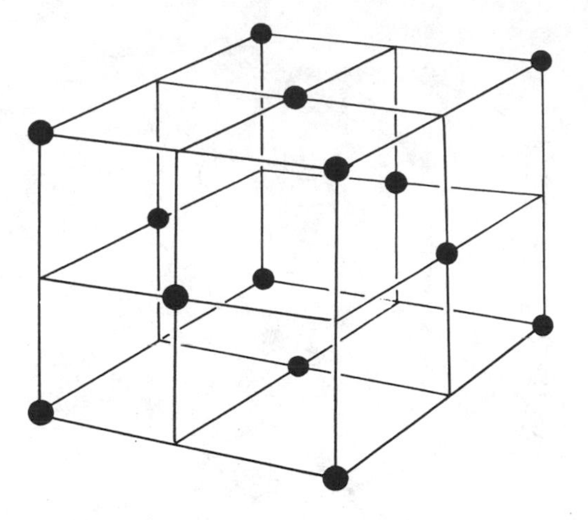

Figure 23. The face-centered cubic lattice of "dry ice." Each unit is a molecule of carbon dioxide.

A different variety of cubical lattice is shown in Figure 24: the *body-centered cubic* (note the unit in the center of the cube). This is a crystal of metallic sodium. Its units are sodium atoms.

Sodium chloride, or ordinary table salt, also has a cubical lattice of atoms (Figure 25), but they are atoms that have become ionized. Sodium has only one electron in its outer shell. Chlorine has seven electrons in its outer shell, but there is room for eight; in a manner of

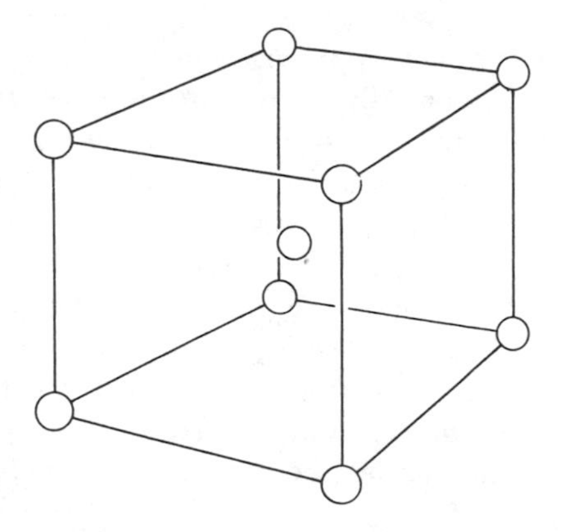

Figure 24. The body-centered cubic lattice of metallic sodium. Each unit is a sodium atom.

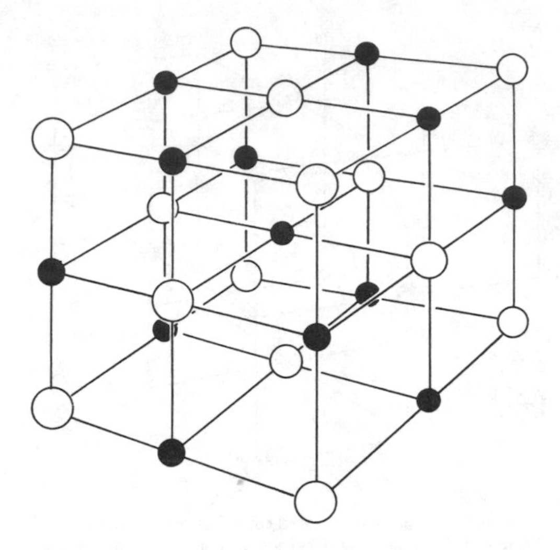

Figure 25. The cubic lattice of table salt. Sodium ions (shown black) alternate with chlorine ions (white).

speaking, there is an empty space into which one electron can fit. When the two atoms get together, the lonely sodium electron leaps into the vacant chlorine space to form a strong, stable molecule of sodium chloride. Because each atom in this lattice has either lost or gained an electron, it has either a positive or negative charge. As mentioned earlier, such an atom is an ion. The units of this crystal are ions.

The lattice of a crystal has a strong influence on the larger forms in which the substance normally occurs. In the case of table salt, the various planes in the cubic lattice form planes along which salt tends to cleave easily. If you examine table salt carefully through an ordinary magnifying glass (of course a microscope is even better) you will see that the grains are actually tiny little cubes. You are not seeing the basic cubical unit shown in the diagram—this is below the magnification level of even the most powerful microscope—but you *are* seeing tiny little crystals of salt that have acquired a cubical structure because of the cubical nature of the salt lattice.

You must not think that just because lattice structures are below microscopic range they are no more than theoretical constructions that physicists have not been able to observe. At one time this was true, but now there are many techniques for actually "seeing" structures much too small to be seen by visible light. As early as 1912 the German physicist Max von Laue developed a technique for observing lattice structure by means of X-rays. More recently, greater precision of detail has been obtained by shooting electrons, ions, and even neutrons through crystals. The cover of *Scientific American* for June 1957 was a striking color photograph showing the arrangement of individual atoms in the lattice of tungsten. The photograph was taken with a new instrument called the field ion microscope, which enlarged the lattice by some two million diameters! So you see, these structures are no longer mathematical guesses; they have come within the range of relatively simple, direct observation.

All three of the cubical lattices just described are symmetric in the sense that we have been using the word; that is, they are superposable on their mirror images. In addition, the three lattices have many other types of symmetry that are studied by crystallographers. For example, they have various kinds of rotational symmetry. This means that if they are rotated in certain ways, about certain axes, the lattice, after the rotation has been made, is exactly the same as it was before. For example, if an axis is passed through a cube as shown in Figure 26, you can rotate the cube into four different positions that are exactly alike, point for point, so far as all features of the cube are concerned. Such an axis is called a fourfold axis of symmetry. It is easy to see that a cube has three such axes.

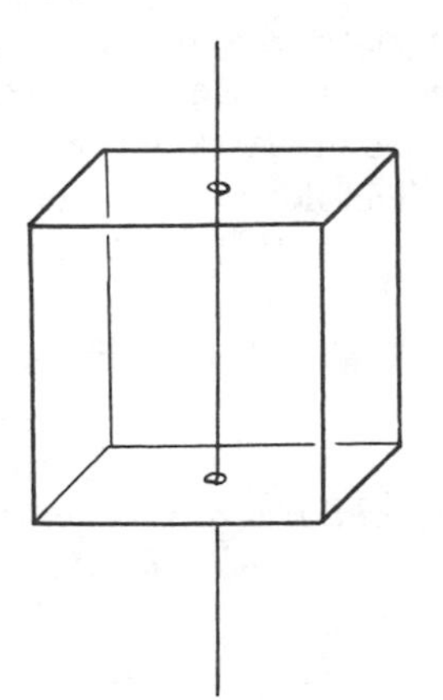

Figure 26. One of the cube's three fourfold axes of symmetry.

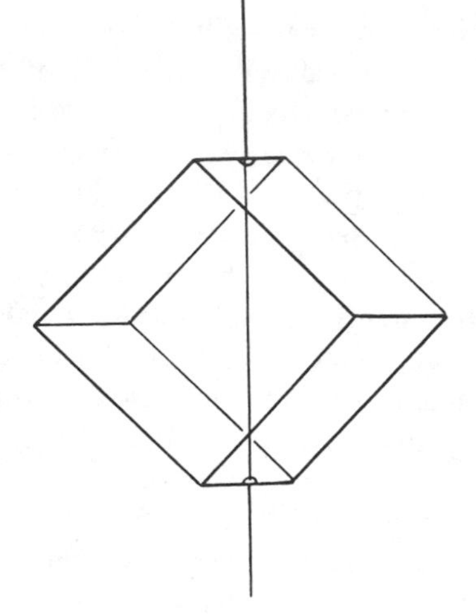

Figure 27. One of the cube's six twofold axes of symmetry.

If an axis is passed through a cube as shown in Figure 27 the cube can be rotated into *two* positions that are exactly alike. Such an axis is called a twofold axis of symmetry. The cube has six such axes.

Crystals may have axes of twofold, threefold, fourfold, and sixfold symmetry. It is not possible for a lattice to have fivefold symmetry. Triangles, squares, and regular hexagons can be used for tiling a floor, but if you tried to tile it with regular pentagons you would run into trouble. For a similar reason pentagonal forms are never found in three-dimensional crystals. They are common in the living world—most flowers (e.g., the primrose) and some animals (e.g., the starfish) exhibit pentagonal symmetry—but you will never find a pentagonal crystal. The underlying lattice structure of crystalline substances, by iron laws of geometry, cannot have a fivefold axis of symmetry.

As we have seen, a cube has axes of both twofold and fourfold symmetry. It does not have a sixfold axis of symmetry. Does it have a threefold axis? Most people are dumbfounded when first told that it has four such axes.

EXERCISE 8: *Find the cube's four threefold axes of symmetry. In other words, find four axes such that, when the cube is rotated around each axis, it can be brought into three positions, no more, no less, which, point for point, are exactly alike.*

All these examples of rotational symmetry are called performable operations for the simple reason that they can actually be performed. Reflection symmetry is called a nonperformable operation because there is no way it can actually be performed with a solid object. As we have seen, a two-dimensional object on a plane can be reflected by picking it up and turning it over, but of course to do this we have to carry the two-dimensional object into 3-space. In the same way we could reflect or "turn over" any solid object if we had some way of carrying it through a higher space. Since we have no way of doing this with an actual object, crystallographers speak of the reflection operation as nonperformable. There are other types of nonperformable symmetry operations, but already we have spent more time than necessary on these operations. The subject of crystal symmetry is a complicated, absorbing topic about which enormous books have been written; we must resist the temptation to go into more detail. This is not a book about symmetry in general. We are concerned with crystals only in respect to their reflection symmetry; that is, whether they possess a plane of symmetry and are therefore superposable on their mirror images.

Many minerals are found in large irregular lumps that give only the faintest indication, if at all, of their underlying crystalline structure. A happy exception is the diamond, a form of crystalline carbon. It usually is found as a single crystal, often of great regularity. Its underlying lattice, cubical in form, permits the diamond to take a variety of crystal shapes. The most common form, shown in Figure 28,

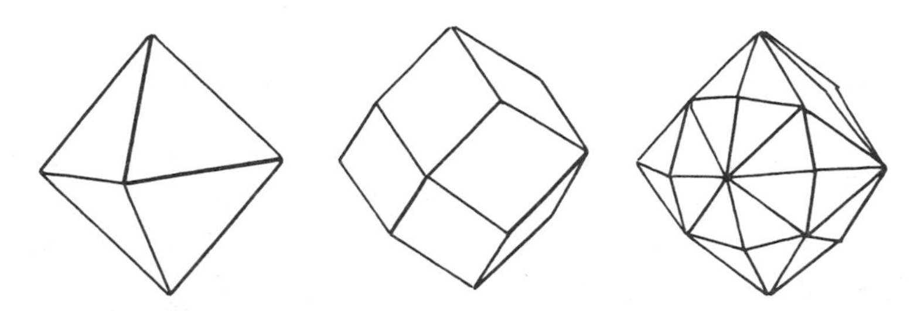

Figure 28. Three natural crystal forms of diamond: octahedron, rhombic dodecahedron, and hexakis octahedron.

left, is known to geometers as an octahedron, or eight-sided figure. Note that each face is an equilateral triangle. A solid figure such as this, composed of plane faces, is called a polyhedron. If it can be turned so that each face will rest flat on a table top, it is a convex polyhedron. When every edge of a convex polyhedron has the same length as every other edge, and every angle is the same as every other angle, it is called a regular polyhedron.

There are exactly five regular convex polyhedrons: the tetrahedron, hexahedron (cube), octahedron, dodecahedron, and icosahedron. Sometimes they are called the five Platonic solids because Plato had some interesting things to say about them. They turn up in all sorts of unexpected places in nature; recently it was discovered that certain viruses have shapes like tetrahedrons, dodecahedrons, and icosahedrons.

The rhombic dodecahedron (Figure 28, center) and the hexakis octahedron (Figure 28, right) are two other striking crystal forms in which diamonds are sometimes found. All three of these crystals are symmetric, each possessing many planes and axes of symmetry, each deriving its structure from the underlying structure of the crystal's lattice. The diamond is the lattice structure assumed by carbon when it undergoes great pressure. So tightly packed are the atoms that it is almost impossible to force them closer together; this is precisely why diamond is the hardest natural substance known. In looser lattice form, carbon becomes graphite (such as used in pencil lead), and when the lattice structure is lost entirely, the result is ordinary charcoal or soot. The difference between the black soot on the inner walls of a chimney and the diamond sparkling on a girl's finger is no more than a difference in the pattern taken by carbon atoms!

A common crystalline form, almost as simple as the cube, is the rhombohedron shown in Figure 29. Its six faces are exactly alike, each a rhombus, so that every edge of the solid has the same length. It is as though you took a cube and pushed it out of shape by applying pressure on two directly opposite corners. Large crystals of mineral calcite (calcium carbonate) are often found in this form; also

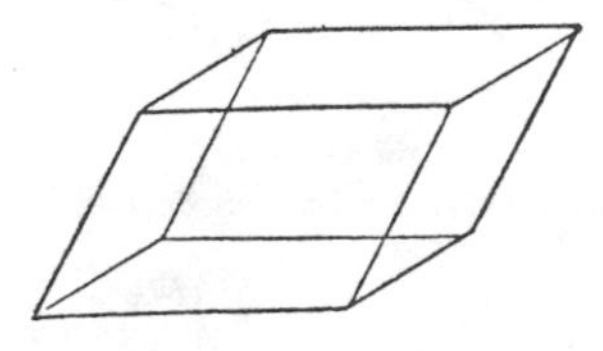

Figure 29. The rhombohedron.

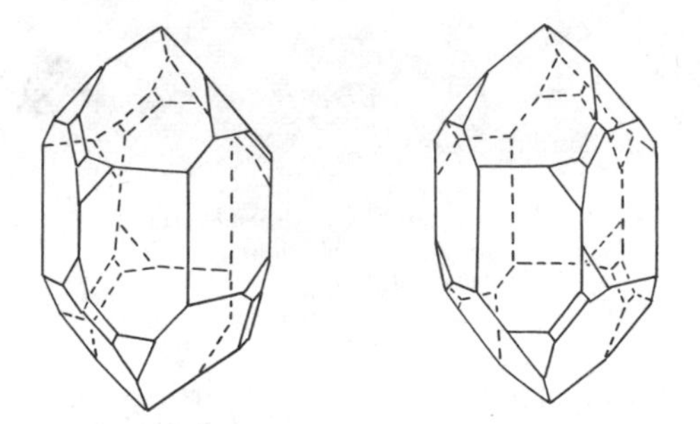

Figure 30. Left-handed and right-handed quartz crystals.

the crystals of sodium nitrate. Can you visualize the shape clearly enough to decide whether it is symmetric or asymmetric?

EXERCISE 9: *Without making a cardboard model, see if you can find one or more planes of symmetry in the rhombohedron. Of course if you find only one, the figure will be symmetric and superposable on its reflection.*

Some crystals found in nature are mirror symmetric in their lattice structure, some are not. Quartz, the most common of minerals, has an asymmetric lattice which is not superposable on its mirror image. Quartz is the compound silicon dioxide, or silica. Its lattice has a helical structure made up of silicon atoms linked with twice as many atoms of oxygen. Because its helices can twist either right or left, quartz has two enantiomorphic forms. In nature, it takes an enormous variety of shapes, which seldom reflect the asymmetric character of its lattice, but on rare occasions an asymmetric quartz crystal like the one shown in Figure 30 is found. The picture shows such a crystal in its two mirror-image forms.

A beam of light normally vibrates back and forth along all planes that pass through the beam's axis. But when light goes through certain crystals such as Iceland spar (a transparent form of the mineral calcite) the crystalline lattice of the mineral allows only a certain plane of light to go through. Light of this sort, undulating along a single plane, is called polarized light. When a plane of polarized light is sent through transparent quartz, the asymmetry of the quartz lattice causes the plane of light to twist sharply in either a clockwise or counterclockwise direction. This provides a simple method by which

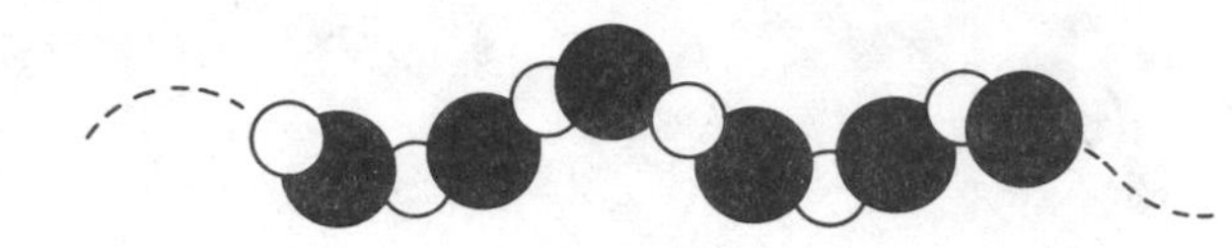

Figure 31. Helical lattice of cinnabar. Mercury atoms (black) alternate with sulfur atoms (white).

the left–right symmetry of many crystal lattices can be tested. Cinnabar (mercuric sulfide), a reddish ore which is the chief source of mercury, will rotate a plane of polarized light to a much greater degree than quartz. Its asymmetric crystal lattice consists of helical chains of alternating mercury and sulfur atoms that twist either right or left in the manner shown in Figure 31.

Moving down the ladder to the third rung, the molecule, a question arises: do molecules themselves, considered as individual units, quite apart from any crystal lattice in which they may be embedded, always have a symmetric structure? If so, then whenever that compound is found in nature or created in a laboratory its molecules will always be the same and the compound will always have the same properties. But if certain molecules are an *asymmetric* structure of atoms, it might be possible to find, or create in the laboratory, two quite different forms of exactly the same compound. One form would contain only "right-handed molecules," the other, only "left-handed molecules." The two substances would be identical in all respects except that the molecules of one would be mirror images of the molecules of the other.

There *are* such molecules. They are called stereoisomers, and the dramatic story of their discovery will be told in the next chapter.

12. MOLECULES

The story of the discovery of left- and right-handed molecules begins in France in the early part of the nineteenth century. Jean Baptiste Biot, a world-renowned French physicist and chemist, had discovered that quartz crystals have the power of twisting a plane of polarized light. A substance capable of doing this is said to be optically active. As we learned in the previous chapter, large quartz crystals are sometimes found in nature in asymmetric forms. It was easy for Biot to determine that if such a crystal rotated a plane of polarized light clockwise, a mirror-image crystal rotated the plane the other way. In addition, he found that if quartz crystals are dissolved in a solution, the solution does *not* twist polarized light. It is optically *inactive.* How can this be explained? Very simply. The twisting ability of quartz must arise not from asymmetry *within* its molecules but from some sort of larger asymmetric structure formed by the molecules themselves whenever quartz is crystallized. This larger structure is, of course, the asymmetric lattice of the quartz crystal.

Biot made another discovery that was not so easy to understand. He found that solutions of certain organic compounds such as sugar and tartaric acid, substances obtained from living things, are also optically active! Why the exclamation mark? Because here, in a solution, there is no crystalline lattice available for twisting polarized light. Ergo, the twist must come from some type of asymmetry within the structure of each individual molecule. Biot had no way of proving this, but it seemed a reasonable hunch.

Biot's work on the optical activity of organic substances, and the guess he made about it, fascinated a young French chemist named Louis Pasteur. Many years later Pasteur became world famous for his great contributions to medical science, but at this time he was in his early twenties, just beginning his career.

Pasteur knew that tartaric acid, a compound found in grapes and certain other fruits, always rotated polarized light a certain way. He also knew that there was another form of tartaric acid, called racemic acid, that was optically inactive. Chemists had found that the two substances were exactly alike in all their chemical properties except one: the ability to rotate polarized light. Tartaric acid twisted the light, racemic acid did not. Here was a curious situation indeed! How can two things be exactly alike in all respects, yet differ in the way they transmit light? Pasteur could imagine only one explanation. Biot must be right. There must be some sort of left–right difference in the structure of the molecules.

Acting on this assumption, Pasteur began an intensive study of the crystal forms of tartaric and racemic acid. He found that the crystals of tartaric acid, examined carefully under a microscope, are asymmetrical; moreover, all the crystals are asymmetrical in the same way. They have the same handedness. But the crystals of racemic acid are an equal mixture of left- and right-handed crystals. Half the crystals are identical with the crystals of tartaric acid, half are enantiomorphic forms (see Figure 32).

It is not hard to guess what Pasteur did next. With great care and patience, using tiny tools that he could observe through the microscope, he separated the crystals of one handedness from the crystals of opposite handedness. When he prepared a solution of one type of crystal he found it identical in all respects with the tartaric acid extracted from grapes. It rotated a plane of polarized light in the same direction as did the naturally occurring tartaric acid. When he prepared a similar solution using the other type of crystal, he also obtained optically active tartaric acid, but with an all-important difference: it rotated polarized light the other way.

"Pasteur was so overcome with emotion," wrote René Dubos in his *Pasteur and Modern Science* (Anchor, 1960), "that he rushed from the laboratory, and, meeting one of the chemistry assistants in the hall, embraced him, exclaiming, 'I have just made a great discovery . . . I am so happy that I am shaking all over and am unable to set my eyes again to the polarimeter!' " As Dubos points out, to appreciate the greatness of Pasteur's discovery we must remember that his laboratory was small and primitive, and that he had been working in it for only two years. He had to prepare all his own chemicals, build all his own equipment. "He had no assistance," writes Dubos, "only the encouragement of his teachers and school friends, and faith in his destiny."

Pasteur's discovery strongly confirmed Biot's hunch about the

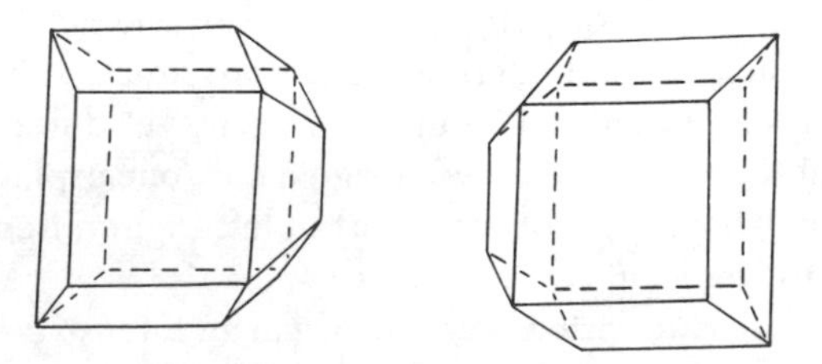

Figure 32. Tartaric acid crystals of opposite handedness.

asymmetry of certain molecules. When the old man heard about the young man's discovery, he immediately sent for Pasteur and asked him to repeat, under his own eyes, the experiment Pasteur had performed with tartaric and racemic acids. To make sure there were no mistakes, Biot insisted on providing his own racemic acid. After the solution had evaporated and formed crystals, he watched over Pasteur's shoulder while the young chemist separated the tiny crystals into right- and left-handed forms. Biot insisted on personally preparing the two solutions and examining them through the polarimeter to see how each twisted the light. He chose first the "more interesting" solution, Pasteur later wrote: the solution that represented the "new" form of tartaric acid not previously known.

"Without having to make a reading," Pasteur wrote (I quote from Dubos's book), "Biot recognized that there was a strong levo-rotation [rotation to the left]. Then the illustrious old man, who was visibly moved, seized me by the hand, and said, 'My dear son, I have loved science so deeply that this stirs my heart.' "

It was Pasteur's first great experiment, an experiment that established beyond doubt that molecules were capable of existing in enantiomorphic, mirror-image forms.

Pasteur's second great discovery in this field, ten years later, was the discovery that when a certain type of plant mold was allowed to grow in a solution of racemic acid, the solution became optically active. A series of experiments established that the mold destroyed only the molecules of a certain handedness, but left the mirror-image molecules undisturbed. Evidently some type of asymmetry in the organic substances of the plant mold caused the mold to act on only one type of tartaric acid molecule. In his previous experiment Pasteur himself had separated the two kinds of molecules; here was a new and novel method of doing the same thing.

"The asymmetric living organism," Pasteur wrote, "selects for its nutriment that particular form of tartaric acid which suits its needs—the form, doubtless, which in some way fits its own asymmetry—and leaves the opposite form either wholly, or for the most part, untouched. The asymmetric micro-organism, therefore, exhibits a power which no symmetric chemical substance, such as our ordinary oxidizing agents, and no symmetric form of energy, such as heat, can ever possess: it distinguishes between enantiomorphs. Asymmetric agents can alone display selective action in dealing with enantiomorphs."

As Dubos brings out clearly in his excellent little book, Pasteur thought very deeply about the implications of these experiments. He knew that most organic substances found in living things were optically active. In contrast, solutions of compounds from the nonliving world were invariably inactive. Pasteur decided that only living things could produce a compound of asymmetric molecules that all went the "same way." He had found two methods of forming such compounds in his laboratory, but both methods involved a living agent: in one case, a mold; in the other case, the agent was Pasteur himself, who divided the molecules by dividing the crystals that they had formed.

Pasteur became convinced (and he was right) that only in living tissues are to be found asymmetric substances composed of just one type of asymmetric molecule. This was, he believed, the only "well-marked line of demarcation that can at present be drawn between the chemistry of dead matter and the chemistry of living matter."

"Non-living, symmetric forces," Pasteur wrote, "acting on symmetric atoms or molecules, cannot produce asymmetry, since the simultaneous production of two opposite asymmetric halves is equivalent to the production of a symmetric whole, whether the two asymmetric halves be actually united in the same molecule . . . or whether they exist as separate molecules, as in the left and right constituents of racemic acid. In any case, the symmetry of the whole is proved by its optical inactivity."

In a moving letter to a friend in 1851, Pasteur wrote (I quote once more from Dubos): "I am on the verge of mysteries, and the veil which covers them is getting thinner and thinner. The night seems to me too long." By that last sentence Pasteur meant no more than that he could hardly bear the night's interruption of his work, so eager was he to get back to his laboratory.

Pasteur had no way of knowing the exact geometrical nature of the asymmetry that caused one molecule to differ from its mirror image, but that such an asymmetry existed he had no doubt. "The molecular structures of the two tartaric acids are asymmetric," he wrote, "and, on the other hand, they are rigorously the same, with the sole difference of showing asymmetry in opposite senses. Are the atoms of the right acid grouped on the spirals of a right-handed helix, or placed on the solid angles of an irregular tetrahedron, or disposed according to some particular asymmetric grouping or other? We cannot answer these questions. But it cannot be a subject of doubt that there exists an arrangement of the atoms in an asymmetric order having a non-superposable image. It is not less certain that the atoms of the left acid realize precisely the asymmetric grouping which is the inverse of this."

In the 1860s a number of chemists suggested that the optical asymmetry of organic compounds might arise from a tetrahedral carbon atom, but it was not until 1874 that this notion was presented as a systematic theory. Biot was then no longer living, and Pasteur was fifty-two. As is so often the case in the history of science, the correct theory was advanced at the same time, and independently, by two men: in this case, one a young Frenchman, Joseph Achille Le Bel, the other a young Dutchman, Jacobus Henricus van't Hoff. Both men suggested that the carbon atom in a carbon compound is situated in the center of a tetrahedral structure, united by chemical bonds with four other atoms located at the four corners of the tetrahedron. The carbon atom has room for eight electrons in its outer shell, but contains only four. Thus it has, so to speak, four empty places into which can be fitted electrons from the outer shells of four other atoms. If no two of these four attached atoms are alike, Le Bel and van't Hoff reasoned, the tetrahedral structure will be asymmetrical and non-superposable on its mirror image.

Isaac Asimov, in a section on the carbon atom in volume 2 of his *Intelligent Man's Guide to Science* (Basic, 1960), suggested an easy way to build a model of a tetrahedral carbon compound. Let a marshmallow represent the central carbon atom. With four toothpicks attach four black olives to the marshmallow to form the tetrahedral structure shown in Figure 33. The black olives represent four other atoms, all of the same element. For example, if each black olive is a hydrogen atom, then you have before you a model of methane, or marsh gas. The formula for methane is CH_4. This means that four atoms of hydrogen are linked by chemical bonds to one atom of carbon, forming a single molecule of methane. The carbon atom,

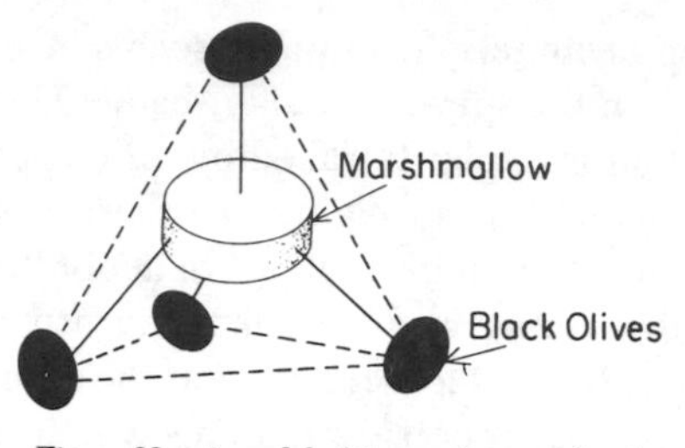

Figure 33. A model of the methane molecule.

you recall, has room in its outer shell for four more electrons. Each hydrogen atom has one electron, so it is easy for four hydrogen atoms to attach themselves to one atom of carbon. When hydrogen and carbon combine, the compound is called a hydrocarbon. The methane molecule is one of the simplest of all hydrocarbon molecules. In Pasteur's day it was diagramed (as it still is today) by using dashes to represent the chemical bonds connecting the four H's (hydrogen atoms) with the one C (carbon atom):

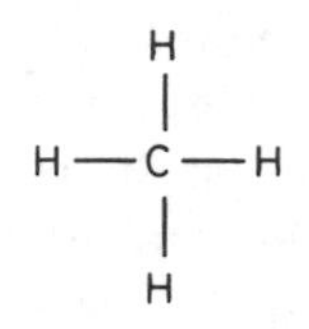

The diagram is, of course, confined to a plane. Le Bel and van't Hoff said to themselves: Suppose we think of this structure as a stable configuration in 3-space. What sort of structure would it be? The tetrahedron, simplest of the five Platonic solids mentioned in the previous chapter, immediately came to mind because it would place each hydrogen atom at the same distance from every other hydrogen atom. The carbon atom (marshmallow) is in the center, an equal distance from each hydrogen atom. Such a molecule is clearly symmetrical. In fact, it has many planes of left–right symmetry. It can be superposed on its mirror image.

Suppose, now, that we take away one of the black olives and substitute a green olive. Is the model still symmetrical? Yes, there are still three planes of symmetry, each bisecting the green olive. One such plane is shown in Figure 34. The model is still superposable on its

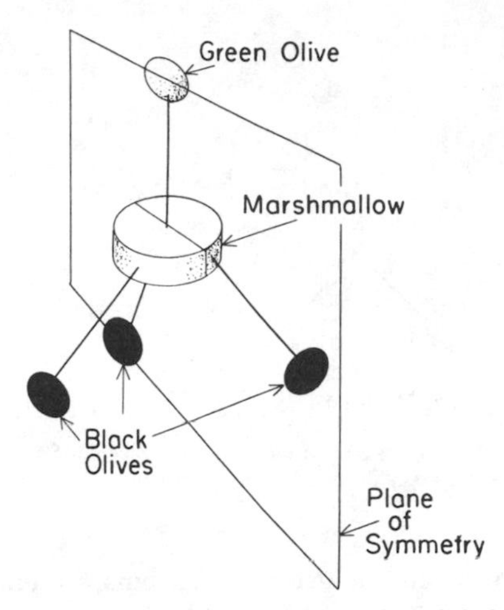

Figure 34. A model of the molecule of wood alcohol, showing one of its three planes of symmetry.

mirror image. Methanol, or wood alcohol, the simplest of the alcohols, is an example of such a structure. Its formula, CH_3OH, is diagramed like this:

```
      H
      |
H — C — O H
      |
      H
```

Take away another black olive, this time replacing it with a cherry. Have you destroyed the model's symmetry? At first thought you might say yes, but if you consider it more carefully you will soon realize that the answer is no. The model is still symmetrical.

EXERCISE 10: *Show how to pass a plane of symmetry through the model (Figure 35), proving that it is superposable on its mirror image.*

An example of this type of structure is found in ethyl alcohol, or grain alcohol, which has the formula C_2H_5OH. In its diagram below you see that the carbon atom is attached to two atoms of hydrogen that are, of course, alike; the other two links are with groups of atoms that are not alike.

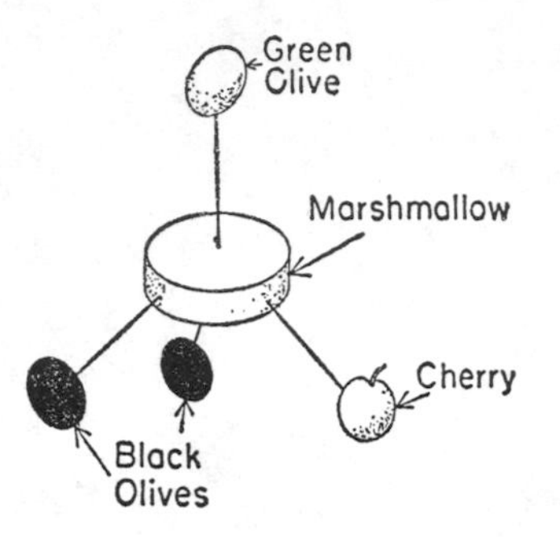

Figure 35. A model of the molecule of grain alcohol. Is it symmetrical?

If at least two atoms or groups of atoms attached to the central carbon atom are alike, the molecule is symmetrical. But if you take

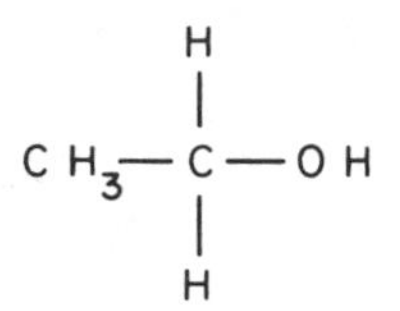

away a third green olive and substitute a cocktail onion, the symmetry is destroyed at last (Figure 36). There is now *no* plane of symmetry. No matter how you turn this model in 3-space, you cannot make it coincide with its mirror image.

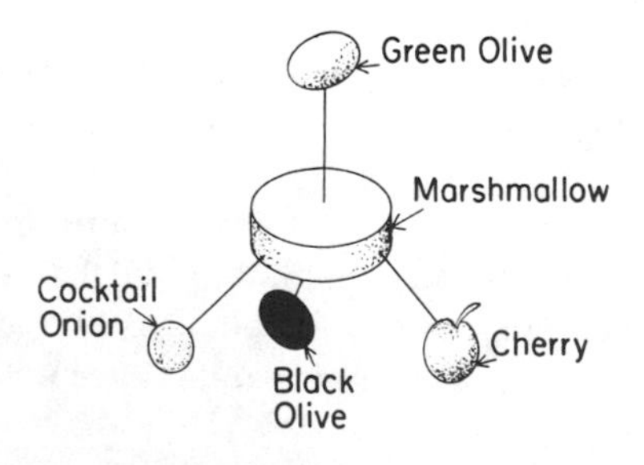

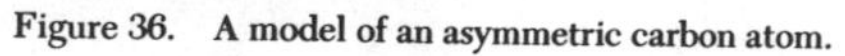

Figure 36. A model of an asymmetric carbon atom.

An example of this type of molecule is provided by one type of amyl alcohol, diagramed as follows:

$$\begin{array}{ccc} & C_2H_5 & \\ & | & \\ CH_3 - & C & - H \\ & | & \\ & CH_2OH & \end{array}$$

As you see, each of the four structures to which the central carbon atom is linked is different. Whenever this is the case, the carbon atom is called an asymmetric carbon atom. Of course the carbon atom itself is not asymmetrical; it is asymmetrical only in the sense that it is linked with four other atoms or groups of atoms in such a way that an asymmetric 3-space structure results. Any molecule containing one or more asymmetric carbon atoms is usually asymmetrical. The exceptions occur when asymmetric atoms of opposite handedness balance each other in much the same way that our left ear balances our right. An example is provided by a fourth type of tartaric acid called mesotartaric.

The diagrams in Figure 37 make clear how mesotartaric acid differs from the other three forms. A right-handed tartaric molecule contains two asymmetric carbon atoms, both right-handed. A left-handed tartaric molecule contains two asymmetric carbon atoms, both left-handed. Racemic tartaric acid is a mixture of equal parts of left- and right-handed molecules. It is said to be *externally compensated.* It is optically inactive because the number of molecules twisting polarized light one way is balanced by the number that twist it the other way. Mesotartaric acid is also optically inactive, but for a slightly different reason: each of its molecules is made up of a right-handed carbon atom attached to a left-handed one. Such a molecule is said to be *internally compensated.* It is bilaterally symmetric in the same way that your head is symmetrical in spite of your asymmetric ears.

To sum up, a molecule may contain asymmetric atoms and still, in an overall way, be symmetrical. A molecule may contain no asymmetric atoms and still have an overall structure that is asymmetrical.

Every compound made up of asymmetric molecules has a right-handed form and a left-handed form. Some such compounds have racemic forms in which left- and right-handed molecules are mixed. In rare cases, mixed molecules can link to make a mesoform.

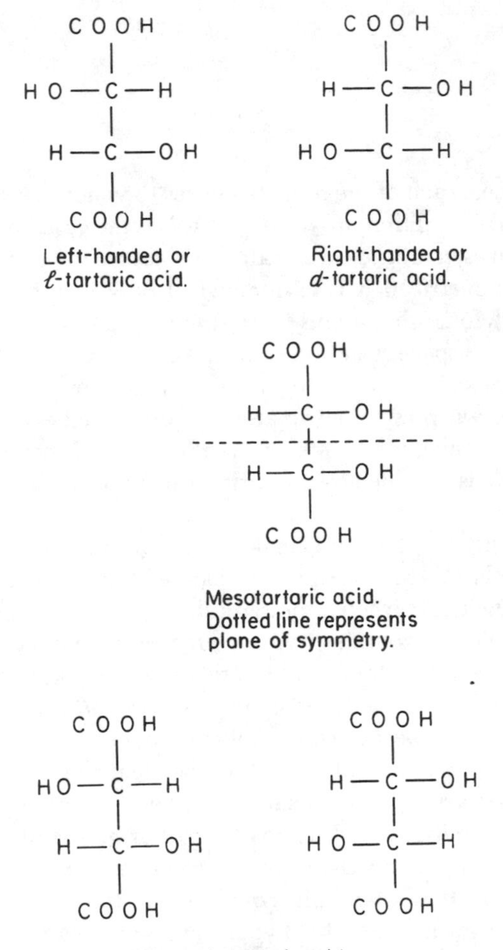

Figure 37. Four kinds of tartaric acid.

Every asymmetric molecule can take either of two enantiomorphic forms. A compound made up of molecules of one handedness will, because of its asymmetric electromagnetic field, rotate a plane of polarized light in one direction. A compound made up of the same molecules, but of opposite handedness, will rotate polarized light by exactly the same degree in the reverse direction. Any substance that rotates polarized light clockwise (as you face the substance, with the substance between you and the light source) is said to be dextrorotary. If it rotates the light counterclockwise it is levorotary. (*Dexter* and *laevus* are Latin for right and left.) The handedness of an optically active substance is indicated by prefixing *dextro-* or *levo-* to its name, or simply, *d* or *l*. Thus, right-handed tartaric acid is called dextrotartaric or *d*-tartaric, left-handed tartaric acid is called levotartaric or *l*-tartaric.

When van't Hoff and Le Bel independently suggested an asymmetric tetrahedral structure as an explanation for optical activity, many scientists scoffed at the notion. One of van't Hoff's colleagues actually dismissed it as "miserable speculative philosophy." It was not long, however, until evidence supporting the theory became overwhelming. We know today that almost every substance found in living organisms is a carbon compound possessing a basic asymmetry, or "chirality" as chemists like to call it, using a term coined by Kelvin.

You must not think that there are perfect little tetrahedrons inside such compounds. The tetrahedral model is only a rough way of picturing the structure of chemical bonds which can be described precisely only by the mathematical equations of modern chemical theory. For our purposes, however, it is accurate enough. Some of the fascinating details and implications of the asymmetric carbon atom will form the content of the next chapter.

13. CARBON

Biochemists (chemists who study the processes of living things) find it difficult to imagine any kind of life—except possibly a sluggish, low-grade form—that does not require tens of thousands of different kinds of tissues, each designed to do a highly specialized job. Think of the complexity of the eye alone, only one of the body's many organs. Special compounds have to be synthesized by the body to provide every component part: the lens, the muscles that change the lens's shape, the muscles that open and close the pupil, the iris, the layers of the cornea, the liquids that fill the chambers, the retina, the choroid, the sclerotic, the optic nerve, the blood vessels. Every part requires enormously complicated substances that have the necessary properties to do precisely what they are supposed to do.

Billions of such specialized tissues are essential to earth's living forms. It is hard to see how evolution could have developed such tissues without the help of carbon, an element that surpasses all others in its ability to form a virtually unlimited variety of compounds, each with its own unique set of properties. There are more than twice as many known carbon compounds as all other known compounds put together. The tissues of every living thing on the face of the earth, from a submicroscopic virus to an elephant, are made of substances containing carbon. Some biochemists go so far as to define life itself as one of the complex properties of carbon compounds.

How does carbon manage to be such a versatile, adaptable element? The answer is that it is a great "joiner." Because its outer shell has space for four more electrons, it can link itself with other carbon atoms to make a chain of indefinite length, and each link in the chain (each carbon atom) will have two spots, so to speak, at which other atoms or groups of atoms can be attached like charms on a charm bracelet. The chain can be a simple one, with two ends like a piece of string. It can fork, like a branching road, and have many loose ends. It can join ends to form closed loops or rings. Rings and chains can be combined in the same molecule. Figure 38 shows only a few of the simpler of millions of patterns that carbon atoms can form by linking together in different ways. Each dash represents a chemical bond to which another atom or group of atoms can be attached to form what chemists call side chains.

When two molecules are exactly alike in the number and kinds of atoms they contain, but differ in the way the atoms are joined together, they are said to be isomers (from the Greek, meaning "of like parts"). Think of each molecule as a set of balls of different colors (all

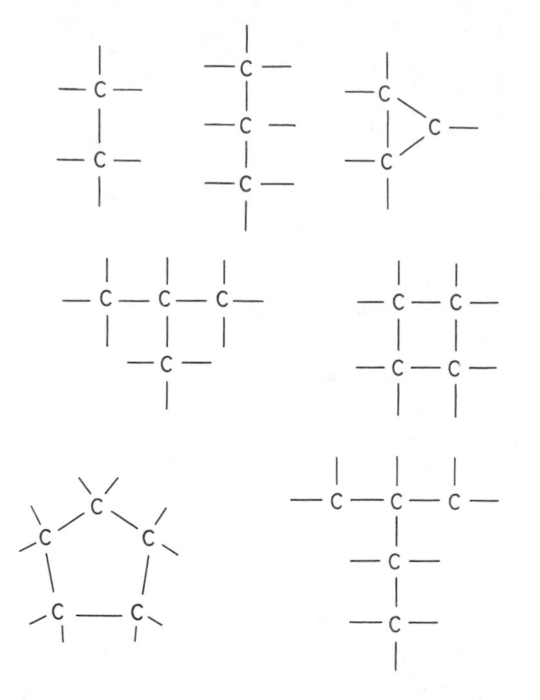

Figure 38. Examples of simple ways in which carbon atoms can link together.

atoms of the same element having the same color) joined to each other by elastic bands. The two molecules have exactly the same number of balls of each color, but they are linked together in different ways. Because of this topological difference in the network by which they are connected, the two isomers may differ in specific gravity, boiling point, and all sorts of other significant properties. A simple example of isomerism is provided by the two topologically distinct ways in which four atoms of carbon and ten atoms of hydrogen can be linked together. If connected as shown in Figure 39, left, the compound is butane; connected as shown in Figure 39, right, it is isobutane.

Is it possible for two molecules to be exactly alike, not only in the number and kinds of their atoms but also in the way their atoms are linked, and still be "different"? Yes; the question was answered in the preceding chapter. One structure can be the mirror image of the other. This form of isomerism is called stereoisomerism. (The prefix

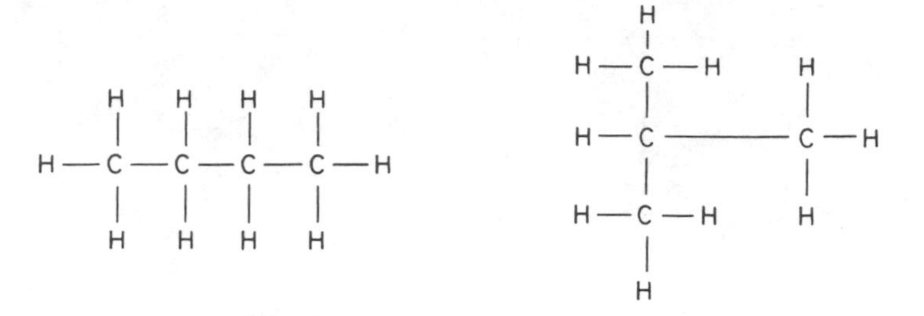

Figure 39. Butane (left) and isobutane (right) have the same atoms, but the atoms are differently linked.

stereo-, from the Greek word for "solid," refers to the fact that stereoisomerism involves structures that must be regarded as three-dimensional, like the tetrahedral models of the carbon molecules discussed in the previous chapter.) Whenever the overall structure of a molecule is asymmetrical, that molecule must have a mirror-image form. In addition, parts of the molecule may be asymmetric, and each part can take either of two mirror-image forms. For example, if a carbon compound contains five asymmetric carbon atoms, each atom can take left- or right-handed forms, thus making possible a large number of stereoisomers. It is not unusual for a giant carbon molecule to have millions of isomers, of which tens of thousands are stereoisomers. Stereoisomerism is a technical, complicated topic, but we need concern ourselves with only one simple fact: every molecule with an asymmetric structure has a stereoisomer that is its exact duplicate in every respect except that it has opposite handedness.

Whenever an asymmetric compound is found in nature, not as the result of a living process, it is always found in racemic form; that is, an equal mixture of left- and right-handed molecules. The reason is easy to understand. The forces of nature—gravity, inertia, and so on—have no bias for right or left. While the compound is being formed, laws of chance dictate that molecules of each handedness will be formed in equal amounts. Even in the laboratory, if stereoisomers are synthesized without applying some type of asymmetry, the result will be a racemic, symmetric mixture that does not twist a plane of polarized light.

Imagine that you have before you a box containing thousands of uncooked alphabet noodles, all shaped like the letter *R*. Because these are solid forms in 3-space, not letters printed on a plane, each *R* has a plane of symmetry and therefore is symmetrical. Suppose now that you dump all the *R*-noodles out on a table, spread them with your hands until each lies flat, then spray them with red paint. Each noodle automatically becomes an asymmetric figure as soon as one of its sides is painted red. Because approximately the same number of noodles have fallen with their left sides up as have fallen with right sides up, as many noodles will be painted red on their left side as on their right. Result: an even mixture of left- and right-handed noodles. Something like this happens when stereoisomers are formed, either in nature or in the laboratory, by any symmetric procedure that does not favor one handedness over the other.

The application of a left–right bias, for the purpose of synthesizing a stereoisomer of a particular handedness, can be made in many different ways. We learned in the previous chapter how Pasteur synthesized both left- and right-handed tartaric acid by dividing the crystals of a racemic mixture into left- and right-handed sets. In this case it was Pasteur's own sense of left and right that applied the asymmetry. We could do essentially the same thing with our "racemic" mixture of left and right *R*-noodles. We simply examine the noodles, one at a time. If an *R* is red on its left side we toss it into one box; if it is red on its right side we toss it into another box. Pasteur also found ways of synthesizing one-handed stereoisomers by exploiting the asymmetric habits of other living things such as bacteria and molds. This could be done with our noodles if we could find some type of organism that would attack and destroy only right-handed noodles. It would leave, of course, a pure collection of left-handed ones.

A third method of synthesizing one-handed stereoisomers, also discovered by Pasteur, makes use of an asymmetric compound that has been previously synthesized or taken from a living organism. A racemic mixture A is combined, say, with a right-handed compound B. The two resulting compounds are not enantiomorphic, because one is a combination of two right-handed substances, the other a combination of opposite-handed substances. Since they are not enantiomorphic, they may differ in some chemical property such as solubility that makes it possible to remove one and leave the other. The final step, removing B from A, leaves a pure, one-handed A.

A crude noodle analogy would be to spread the racemic mixture of *R*-noodles over a table top punctured with thousands of small holes, each hole the shape and size of a noodle and with the form of an unreversed R when viewed from above. All the noodles lie flat, red sides up, but half of them are R-noodles, half are Я-noodles. If we shuffle the noodles about over the surface, taking care that none flip over, only R-noodles will fall through the holes. This will leave on the table a mixture in which Я-noodles predominate. Here the left–right bias is supplied not by the process of shuffling but by the asymmetric structure of the table top. The table symbolizes an asymmetric compound. It can, in chemical reactions, impress its asymmetry, so to speak, on other compounds that are racemic mixtures of left and right forms. The new asymmetric compound can then in turn impress *its* asymmetry on other racemic mixtures, and in this way the total amount of asymmetric molecules is steadily increased. This is important to understand because, as we shall learn later, it was probably in just such a way that a few asymmetric compounds, in the early history of the earth, were able to impress their handedness on almost all the molecules in living things today.

To summarize: Some type of left–right asymmetry—whether originating in the chemist's own sense of left and right or in the substances, forces, or living organisms that play a role in the laboratory procedures—must enter at some point into every method of synthesizing one-handed stereoisomers.

Almost every carbon compound found in living things is a stereoisomer of single handedness that twists polarized light in one direction or the other.[1] A familiar class of such optically active organic compounds is the sweet-tasting carbohydrates called sugars. Most of them are right-handed. Ordinary table sugar, for instance, or sucrose, rotates polarized light to the right. So does grape sugar, a form of glucose. Grape sugar is sometimes called dextrose because of its right-handedness. Fructose, or fruit sugar, on the other hand, rotates polarized light the other way and for that reason is often called levulose. It has exactly the same atoms in its molecule as grape sugar, but the way they are linked together gives it a sweeter taste than dextrose and also makes it less harmful to diabetics than either dextrose or sucrose.

The most complicated, as well as the most numerous, of all the carbon stereoisomers are the proteins. Every living organism on earth contains some type of protein. A human body is believed to contain some hundred thousand different kinds of protein. A single cell of the human body may have within it a thousand different enzymes (necessary aids to a thousand different chemical reactions) and every enzyme is a protein. Most hormones (regulators of growth and activity) are proteins. Not a single part of the body, including bone, blood, muscle, tendons, skin, hair, and fingernails, escapes having some kind of protein in its structure. It was mentioned earlier that many biochemists think that life is not possible without the versatility of carbon. Some biochemists think it is not possible without the versatility of proteins.

The protein molecule is made up of atoms of carbon, hydrogen, oxygen, nitrogen, and usually, though not always, sulfur. It is the largest, most complicated of all molecules. A relatively simple protein molecule contains a thousand or so atoms. Giant protein molecules contain hundreds of thousands of atoms, and supergiant protein molecules have more than a million. Each molecule is composed of distinct subunits called amino acids, joined together like the links of a chain. Giant molecules of this sort, made up of units which in themselves can be considered molecules, are known as polymers.

There are some twenty different varieties of amino acids, all but one (glycine) with an asymmetry of either right or left form. When an amino acid is synthesized in the laboratory it is a racemic mixture of both types of handedness, but in the proteins of living things (with only a few rare exceptions) it is always left-handed. This does not necessarily mean that it will twist plane-polarized light in a counterclockwise direction. The side chains of an amino acid also influence the way in which it twists polarized light. All amino acids found in living things are left-handed in terms of the arrangement of atoms around the carbon atoms; but some of them, because of the structure of their side chains (chains of atoms attached to the carbon atoms), rotate plane-polarized light clockwise.

In addition to the left-handedness of all its amino acid subunits, every protein molecule found in nature has a "backbone" which coils into a helix. This backbone, sometimes called a polypeptide chain, is simply the basic chain of amino acids. Every amino acid has an amine end and an acid end. When opposite ends come together, a water molecule is removed from them by the extraction of one hydrogen atom from the amine end and single atoms of oxygen and hydrogen from the acid end. Electrical forces then weld the ends together in what is called a peptide link. Each left-handed amino acid contributes the same twist to the protein molecule's backbone, just as each asymmetric step in a spiral staircase contributes the same sort of twist to the stairway. As a result, the backbone coils into a helix of the type shown in Figure 40. It is called the alpha helix. Linus Pauling and Robert B. Corey, biochemists at the California Institute of Technology, were the first to discover and name this helical structure. Since their pioneer work in the early 1950s the alpha helix has been found in so many other proteins that most biochemists today think it is characteristic of all giant protein molecules.

Should the alpha helix be called right-, or left-handed? If you look at either end of such a helix you will see it coiling toward you in a leftward or counterclockwise direction. For this reason it could be called a left-handed helix, and in fact it *is* so called by some biochemists when they describe a helix in nature, such as the helix of a climbing plant. (In chapter 7 we discussed briefly this confusion of terminology with respect to the twining plants.) On the other hand, this is the type of helix found on ordinary wood screws, commonly called right-handed. Moreover, in crystalline structures such as quartz and cinnabar, this type of helix will rotate a plane of polarized light to the right, or clockwise. For these reasons biochemists speak of the alpha helix as right-handed. It is confusing, one must admit, to say that left-handed amino acids cause a protein molecule to coil into a right-handed helix. However, it is all a trivial matter of words, and the source of the confusion is easy to understand. The important point is that almost every protein found in living things is now believed to

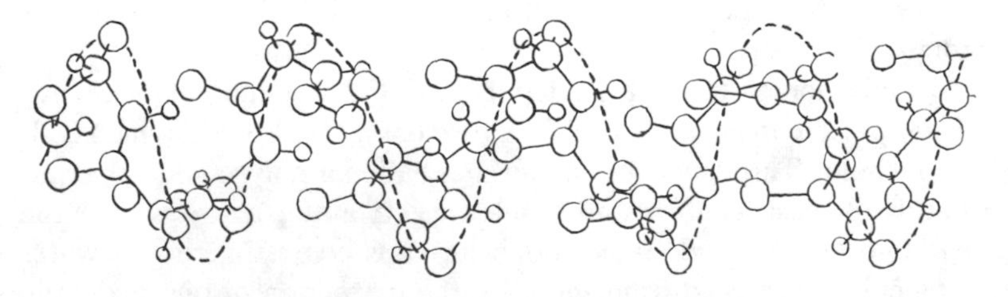

Figure 40. Model of a polypeptide chain, showing the helical structure of its backbone.

have a helical backbone of the same handedness as that of a right-handed corkscrew.

In many structures of the body the alpha helix causes fibrous body tissues to coil in the same right-handed way, producing what has been called "coiled coils." Collagen fibers, for example, found in tendons and other parts of the body, are now thought to have a molecule that consists of three alpha helices twisted together to form a right-handed triple helix. Ten of these triple helices twist together to form a still larger helix. These in turn twist together to form a still larger helix. The "coiled coils" continue to repeat on larger levels until a helical fiber is produced that is large enough to be seen even with an ordinary microscope. Other instances of right-handed coiled coils are found in the fibers of hair, wool, and horn, and in the structure of the flagellum of bacteria (a whiplike appendage that propels a bacterium through a liquid). In the next chapter we shall learn that a right-handed helix is also found in the structure of nucleic acids, carbon compounds even more essential to life as we know it than the proteins.

The number of possible protein molecules, no two alike, is almost infinite. It is infinite in the same sense that the number of different words that can be formed with the twenty-six letters of our alphabet is infinite. This assumes, of course, that there is no limit on the number of letters that a single word may contain and still be called a word. When you consider the fact that the backbone of a protein molecule may contain a thousand or more amino acid subunits, and each unit can be one of twenty different varieties, you realize that the possibilities for different compounds defy the imagination. Of course, it is precisely this unlimited variety that makes protein such an efficient substance to have on hand for the evolutionary construction of machines as complicated as animals—machines in which thousands of specialized tissues have to perform thousands of specialized tasks.

As one would expect, right- and left-handed forms of any organic compound are exactly alike in all chemical properties except those that involve a right–left difference. They have the same specific gravity, melt at the same temperature, freeze at the same temperature, and so on. This is to be expected, because they *are* the same substance, and because the forces acting upon them (heat, gravity, and so on) show no bias toward right or left. Of course, the asymmetry of such a compound reveals itself in many ways. It will rotate a plane of polarized light. It may cause the compound to form crystals that have a certain handedness. And it will cause specific effects when the substance is swallowed by an animal or injected into an animal's bloodstream. Because an animal's body is made up largely of asymmetric compounds, it is easy to understand why stereoisomers of opposite handedness would have different effects on the animal. Lewis Carroll's White Knight, in *Through the Looking-Glass,* sings a song in which the following lines occur:

> And now, if e'er by chance I put
> My fingers into glue,
> Or madly squeeze a right-hand foot
> Into a left-hand shoe. . . .

The last two lines describe a situation similar to what happens when asymmetric compounds react. It is easy to put your foot into a shoe of the same handedness, difficult to squeeze it into a shoe of opposite handedness. For the same reason there often are marked differences in the taste and smell of stereoisomers of opposite handedness. The nerve endings that initiate the processes of tasting and smelling are made of asymmetric substances which react differently when left- or right-handed substances come in contact with them.[2] If an asymmetric substance is swallowed or taken into the bloodstream by injection, it also comes in contact with asymmetric body compounds. Sometimes a stereoisomer of one handedness is digested and used by the body, whereas its mirror-image twin is simply excreted. In other cases, both forms of the stereoisomer are digested and used in the same way by the body, but the rate of digestion is slower for one form than the other.

In still other cases, the body accepts both forms but reacts differently to each. Cigarettes, for example, contain levonicotine, an asymmetric carbon compound in the alkaloid family. (In this sense we can say that our cylindrical cigarettes are all left-handed.) Levonicotine is found in all tobacco plants. But there is a right-handed form of nicotine, dextronicotine, never found in tobacco plants. It has been synthesized and discovered to be much less toxic than levonicotine. Levohyoscyamine strongly dilates the pupil of the eye; dextrohyoscyamine has only a weak effect. Levoadrenaline is twelve times stronger than its mirror image in constricting blood vessels. The reflected form of vitamin C has almost no effect on the body. Thyroxin, the thyroid hormone, is occasionally given to heart patients to lower the amount of cholesterol in their blood. (Cholesterol is a fatty substance believed to play a role in blocking arteries and causing heart attacks.) Thyroxin is an asymmetric amino acid. In natural form it speeds up body reactions, often causing nervousness and loss of weight. A synthetic thyroxin, the mirror image of natural thyroxin, is said to cut down cholesterol just as effectively, but without undesirable side effects.

Almost all of the millions of asymmetric carbon compounds found in living things occur in only one of their two mirror-image forms. (A few do occur in both forms, but never as separate compounds in the same species). Chemists have synthesized only a small number of stereoisomers that are mirror reflections of those naturally occurring in living things. Because most organic substances are obtainable in only one of their two possible forms, little is known about how the human body (or any other organism) would react to the *other* form of the substance.

Before Alice stepped through the mirror into the nonsense world behind the looking glass, she said to her kitten: "How would you like to live in Looking-glass House, Kitty? I wonder if they'd give you milk in there? Perhaps Looking-glass milk isn't good to drink. . . ." Lewis Carroll could hardly have been aware of how profound a question his Alice was raising. It is true that water, which makes up about 85 percent of cow's milk, has a symmetric molecule unaffected by mirror reflection. But milk also contains a number of asymmetric carbon compounds such as fat, lactose (a sugar found only in milk), and various types of proteins. Nobody knows how a mirror image of this mixture we call milk would affect a cat or child who drank it, so no one really knows whether Looking-glass milk is good to drink or not. Chances are it isn't. Of course, a Looking-glass cat would find it as tasty and nourishing as unreflected milk is to an unreflected cat.

W. H. Auden, a great admirer of the Alice books, raises a similar question in his poem *The Age of Anxiety*. A right-handed Irishman, sitting at a New York bar and contemplating his reflection in a mirror, says:

> My deuce, my double, my dear image,
> . . . What flavor has
> That liquor you lift with your left hand . . . ?

Liquor contains grain alcohol, which, as we saw in the last chapter, has a symmetrical molecule. Like the water in milk, it, too, would be unaffected by mirror reversal. But liquor also contains carbon compounds called esters which give it flavor, and most esters are asymmetrical. No one knows what flavor Looking-glass liquor might have, but it is a good bet that it would not taste the same as ordinary liquor unless, of course, it were tasted by a Looking-glass Irishman.

Outside of living things, compounds found in nature are either symmetrical in the way their atoms are linked together, or, if asymmetrical, both right- and left-handed forms are found in equal quantity. Inside living things the reverse is true. Our bodies are saturated with carbon asymmetry, mostly of the left-handed variety. Mirror-reverse the molecules and crystal structure of gold—it remains exactly the same in structure as it was before. Reflect a glass of milk or a shot of whiskey and it is *not* the same. The molecular structure of certain substances in milk and liquor is not superposable on their mirror images. So with a man. Reflect his amino acids—they turn from left-handed to right-handed. Reflect the alpha helices of his proteins—they turn from right-handed to left-handed. Scarcely a molecule of his body, apart from molecules of water, escapes transformation by the mirror into a molecule that goes, as Alice said, the "other way."

Notes

1. When an organism dies, the molecules of certain of its amino acids start to "flip" (change handedness) at a very slow but fairly uniform rate. After many millions of years these amino acids become racemic, containing about an equal proportion of left and right molecules. This "racemization" process now provides a way of dating ancient objects that may some day prove to be more accurate than the familiar carbon 14 method based on the amount of radioactive decay. Because the rate of racemization is affected by moisture and temperature, there is a considerable margin of error, but it has the great advantage of applying to artifacts older than 40,000 years, the limit of dating by the carbon method.

2. There are several theories about how the nose detects odors. The stereochemical theory asserts that the overall shape of a molecule determines how it smells, rather than its vibratory energy levels. This view is strongly supported by the fact that substances that are identical except for handedness usually have different odors. For example, right-carvone smells like spearmint, left-carvone like caraway. It was recently discovered that the difference in the smell of oranges and lemons is caused by differences between right and left forms of limonene.

14. LIVING MOLECULES

In mathematics it is possible to draw sharp, precise lines that divide mathematical entities into two classes. A geometrical structure is either superposable on its mirror image or not superposable. An asymmetric structure is right-handed or left-handed. Every integer is odd or even. There is no integer whose status in this respect is dubious. But in the world itself, except on the subatomic level of quantum theory, dividing lines are almost always fuzzy. Is tar a solid, or a liquid? Is chartreuse yellow, or green? Most physical properties lie on continuums—spectrums that fade imperceptibly from one end to the other. No matter where you bifurcate them, there will be objects so near the dividing line that ordinary language is not precise enough to enable one to say whether the objects belong on one side or the other.

The property of life is on such a continuum.

To prove this we have only to consider the viruses. These are the smallest known biological structures that have the power to "eat" (absorb substances from their environment), grow, and make exact copies of themselves. They are much smaller than bacteria (in fact, some viruses *infect* bacteria). They pass right through a fine porcelain filter. Millions can be put on the head of a pin. Because they are smaller than the wavelength of light, they cannot be seen in an ordinary light microscope, but biochemists have ingenious ways of deducing their structure from what they see when they bombard them with X-rays or beams of elementary particles.

It is true that a crystal can be said to "grow," but it grows in a relatively trivial way. When it is in a solution that contains a compound similar to itself, the compound will be deposited on its surface; the more deposited, the larger the crystal grows. But the virus, like all living things, grows in a more astonishing way. It takes elements from its environment, synthesizes them into compounds not present in the environment, then puts those compounds together to make a complex structure that is a replica of itself. The power of the virus to infect and sometimes kill an organism is due to this ability. It invades the cells of the host organism, where it takes over the cell's machinery, supplying it, so to speak, with new blueprints. It orders the cell to stop making whatever it normally makes and start making the substances needed for putting together copies of the invading virus.

In its ability to replicate (make copies of itself) the virus acts like a living thing. But when removed from living tissues, it crystallizes. These virus crystals often take the form of beautiful regular and semiregular polyhedrons: tetrahedrons, icosahedrons, dodecahedrons, rhombic dodecahedrons, and so on. The virus crystals are totally inert, showing no signs whatever of life. They are as "dead" as a specimen of quartz. But put such a crystal back into the species of plant or animal it is designed to infect—it again springs into deadly action.

The first virus to be discovered, and one of the most studied, is the simple virus that causes the "mosaic disease" in tobacco plants. This virus crystallizes into tiny rods that can be seen in an electron microscope. It has recently been discovered that each rod is actually a right-handed helical structure formed by about 2,000 identical molecules of protein, each molecule containing more than 150 amino acid subunits. The protein molecules coil around a hollow core that runs from one end of the rod to the other. Embedded in the protein (not in the core, as formerly thought) is a single right-handed helical strand of a carbon compound called nucleic acid.

Nucleic acid is not a protein, but like protein it is a polymer: a compound with a giant molecule that consists of smaller molecules linked together in a chain. The subunits, called nucleotides, are made up of atoms of carbon, oxygen, nitrogen, hydrogen, and phosphorus; but where protein has some twenty different amino acid subunits, nucleic acid has only four different nucleotides. Thousands of nucleotides can bond together, as do the amino acid subunits of protein, in an almost endless variety of combinations to form billions of different nucleic acid molecules. Like the amino acids, each nucleotide is asymmetrical and left-handed. Because of this the backbone of a nucleic acid molecule, like the backbone of a protein molecule, has a right-handed helical form.

Nucleic acid comes in two varieties—DNA (deoxyribonucleic acid) and RNA (ribonucleic acid). Every virus consists of a shell of protein enclosing one or more coils of nucleic acid. The tobacco mosaic virus contains only one coil of RNA. Some viruses contain only DNA, others contain both types of nucleic acid. There is little doubt that the nucleic acid, not the protein, kills the host. When a virus attacks a bacterium, the protein part of the virus attaches itself to the outside of the bacterial cell, where it remains while the coil of nucleic acid bores into the cell and starts issuing new orders to the cell's replicating machinery. Soon the cell is turning out copies, not of itself but of the virus. Hundreds of duplicates of the invading virus, complete with protein cells and internal coils of nucleic acid, burst from the cell to invade other cells.

Like protein helices, a right-handed helix of nucleic acid is often twisted into a right-handed helix of larger size. In 1962 biochemists at Yale described the structure of a virus containing a "coiled coiled coil" of nucleic acid. The virus is the T-2 bacteriophage. (A bacteriophage is a virus that infects only bacteria.) It has a head in the shape of what is called a bipyramidal hexagonal prism (see Figure 41). Attached to this head is a protein tail. Inside the head, capable of extending down into the tail, is a single molecule of DNA that exhibits three levels of helicity. The primary helix is the backbone of the DNA molecule. This coils into a secondary helix, which in turn is wound into a tight little spool that fits snugly into the virus's prismatic head. The virus attaches itself by its tail to the host cell. The tail punctures a small hole in the cell's membrane. Presumably, one end of the DNA molecule is pushed into the hole by a contraction of the tail, then the little spool in the head rotates clockwise as the DNA molecule snakes its way through the opening to begin its dirty work.

Not only is nucleic acid found in every virus; in its DNA form it is inside the nucleus of every living cell, from one-celled organisms like

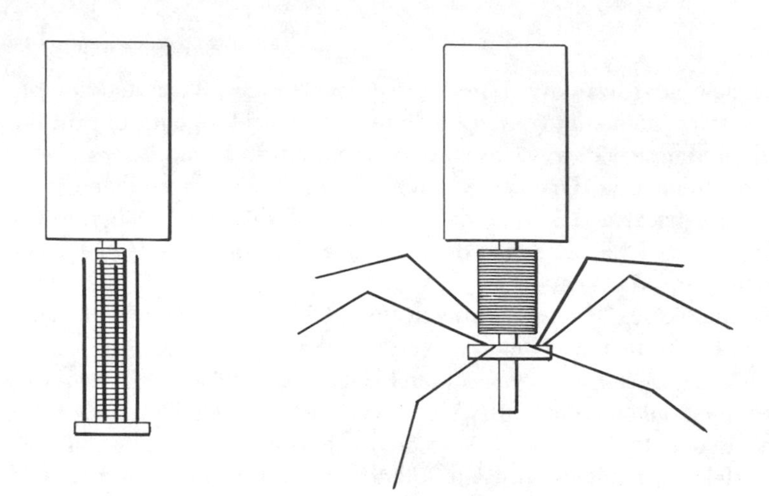

Figure 41. Model of the T-2 bacteriophage in "untriggered" (left) and "triggered" form (right). (Redrawn from an illustration in " The Structure of Viruses," by R. W. Horne, *Scientific American*, January 1963.)

the amoeba to the cells of a human body. The elusive genes—submicroscopic "particles" that carry the organism's genetic code of hereditary information—are not really "things" at all, as formerly thought. They are *regions* along a double-stranded helical molecule of DNA. In every human cell there are forty-six rodlike structures called chromosomes, each containing at least one intertwined pair of right-handed DNA helices. The precise order of the four different nucleotide bases along each coil is the genetic code that tells the cell what to do. (The four bases—adenine, thymine, guanine, and cytosine—are commonly represented by the letters *A, T, G, C.*) Each amino acid is coded by a three-letter combination. With four letters to choose from, there are sixty-four possible three-letter combinations, more than enough to take care of all the amino acids and to "spell out" the exact order in which they must link to produce any given protein. The "gene" is simply a section of the code message—a message that extends from one end of the DNA helix to the other.

Exactly how the message is "punctuated" to mark where a "gene" begins and ends is an aspect of the code that (at the time this is written) has not yet been fully solved. There is growing evidence that in some cases the DNA code of a virus may contain two or even three messages that overlap. Start reading with the first letter of a triplet and you get one message, start with the second or third letter and you may get two other messages. A crude analogy in English is provided by the sequence PIR ATE. Start at the first, second, or third letters and you get three different meanings: PIRATE, IRATE, and RATE.

It has been estimated that if all the DNA helices in one human cell were pulled out straight and placed end to end they would form a thin ribbon about one yard long. Can a repetition of no more than four different symbols, in linear order along this ribbon, carry enough information to govern the growth of an organism as complicated as a human being? It can. There is not the slightest doubt that this yard of ribbon is capable of carrying, in its simple four-symbol code, more than enough information to provide a complete blueprint for the construction, growth, and replication of every individual human.

In 1962 James Dewey Watson, now a biologist at Harvard, and English biologists Francis Harry Compton Crick and Maurice Hugh Frederick Wilkins were given Nobel prizes for their contributions to the discovery of the structure of the DNA helix. It is perhaps the greatest scientific discovery of this century, outranking even those of nuclear physics in its potential impact on history. Twenty years ago the mechanisms of heredity were shrouded in mystery and thought to be enormously complicated. Now suddenly it looks as if the mechanisms may be comparatively simple. Work on cracking the genetic code is proceeding at such dizzying speed that it may soon become possible to control and direct the course of evolution. A full understanding of the code could lead to the creation of synthetic life, to the cure of cancer and other disorders, to an understanding of how the brain stores its memories. So staggering is the biological revolution launched by the discovery of the DNA helix that even Soviet political leaders finally concluded that Trofim Lysenko (the Russian biologist who dismissed modern genetics as a Western bourgeois perversion) was indeed the crank that Western geneticists always said he was.

A nucleic acid molecule in the cell of a plant or animal is a fixed part of that cell. In contrast, the nucleic acid molecule of a virus is a kind of free, wandering set of genes, unattached to any cell but capable of replicating whenever it finds a host cell containing the substances it needs for replicating. Shall we say that the tobacco mosaic virus is "alive"? Most biochemists would. It has two properties that are fundamental in distinguishing living things from the nonliving: it can copy itself and it can mutate. (A mutation is nothing more than a copy that differs in some small way from the original; the difference is passed on to all subsequent copies made by the mutant.) It is estimated that a typical nucleic acid molecule in a cell will make about four million exact replicas of itself before, for one reason or another, it makes a slight error and produces a mutant. That such mutants occur is hardly surprising; the really startling fact is that so *few* occur. Many biochemists today do not hesitate to say that the RNA helix, inside a rod of the tobacco mosaic virus, is of and by itself "alive." They say this because it is the RNA molecule, not its protein shell, that has the power of self-replication and mutation.

We should recognize, of course, that when we debate whether a DNA or RNA helix is "living" or "nonliving" we are tangled in what is essentially a semantic problem. At the level of the nucleic acid molecule the term *life* is simply not precise enough to be useful. *Blue* and *green* are efficient words in ordinary speech: they lose their utility if we try to apply them to a blue green color. *Plant* and *animal* are useful terms, but they fail when one considers simple forms of life that have both plant and animal characteristics. *Bird* and *reptile* are convenient classifications, but where does the *Archaeopteryx* belong? This now-extinct vertebrate is so nearly halfway between reptile and bird that it is a waste of time to argue about whether it is a flying reptile or a reptilian bird.

It is the same with *living* and *nonliving*. Even if we define life as the ability to replicate and mutate, the term has fuzzy boundaries. There is no reason why a computer could not be built someday that would be capable of taking parts from its environment and making replicas of itself, even to mutate. John von Neumann, the great Hungarian mathematician, wrote a famous paper in which he explained how such a machine could, in theory, be constructed. Would we call such a machine alive?

Consider also the fact that there are living organisms, such as worker bees, that are sterile and therefore cannot copy themselves. Yet they are clearly alive. Finally, consider the very real possibility that one of these not-too-distant days a biochemist will synthesize a carbon molecule, something like nucleic acid, that will be capable of making a poor and partial copy of itself. You see, even if self-replication and mutation are made the basic criteria of life, the concept is still blurry. One hears much talk these days about whether later space probes will or will not find life on Mars. A third possibility is seldom considered: the probes may find something on Mars which no one will know whether to call living or not. At the moment, scientists are divided over the question of whether the data sent back by our first probe of Martian soil indicates life, or a chemical reaction not yet fully understood.

The plain fact is, to return to the point made earlier, viruses lie on a continuous spectrum of structural complexity. The spectrum fades back into the nonliving world of crystals and "dead" organic molecules. It fades forward into simple, one-celled forms of plant and animal life. A virus is like the blue green object that can be called either green or blue. It is a structure in the twilight zone—a living-dead thing that our language is not yet rich enough to classify.

Regardless of whether we choose to call a nucleic acid molecule living or not, the fact remains that here at last biochemists have isolated the most essential structure of life as we know it. Pasteur was more right than many of his colleagues suspected when he wrote eloquently of left–right asymmetry as a key to the mystery of life. At the heart of all living cells on earth are right-handed coils of nucleic acid. Dr. Crick, who lives at Cambridge University, has named his house The Golden Helix. This asymmetric structure is surely the master key of life. It carries all the information needed by a living organism to grow into the complicated machine it is, to make copies of itself, and to evolve by the curious procedure of making random copying errors. "If proteins are the principal stuff of life," Dr. Crick wrote in an article on "Nucleic Acids" (*Scientific American*, September 1957), "the nucleic acids are its blueprints—the molecules on which the Secret of Life, if we may speak of such a thing, is written."

We have already raised the question whether on some other planet life can exist apart from carbon compounds. No one knows, of course, but most biochemists think that self-replication and mutation are probably too complex to be carried out by any molecules lacking the enormous range and flexibility of carbon compounds. Silicon comes the closest to carbon in its ability to combine with itself and other elements to form many different compounds, but its chains are relatively short and unstable compared to those of the hydrocarbons (carbon compounds containing hydrogen) that are so essential to life on this planet.

One of the most remarkable and least mentioned characteristics of life as we know it is the ability of an organism to take compounds from its immediate environment, many of which are symmetrical in their molecular structure, and to manufacture asymmetric carbon compounds that are right- or left-handed. Plants, for example, take symmetric inorganic compounds such as water and carbon dioxide and from them manufacture asymmetric starches and sugars. We saw in the previous chapter how riddled the bodies of all living things are with asymmetric carbon molecules, as well as the asymmetric helices of proteins and nucleic acids. Since every asymmetric molecule has a mirror-image stereoisomer, there is no reason why all life on earth could not function just as well if all organisms were suddenly transformed into their mirror images. Of course, if only a single organism, say a man, were reflected, he would probably not be able to survive. His body, with its tens of thousands of asymmetric compounds, would not have the proper handedness for digesting and utilizing the available asymmetric food. But if the molecular structure of *all* living things on earth were reflected—that is, if every stereoisomer in every organism were changed to its mirror twin—the processes of life would continue as before.

How did life on earth get its original left–right twist? Why did organic compounds form the way they did, rather than the other way? Why are all subunits of protein and nucleic acid left-handed? No one knows the answers to these questions because no one knows how life started on the earth. But every day, biochemists are making better and better guesses. The next chapter will give a quick rundown on what present-day science has to say about this fascinating topic.

THE ORIGIN OF LIFE

Almost every nook and cranny of this old earth is teeming with life: life in a fantastic variety of sizes, shapes, colors, sounds, and smells.

How did it all start? Did all living things evolve from one single carbon-containing molecule, or from many different, independently formed molecules? Are such molecules still being formed on earth? No one can claim to know the answers. But for the first time in history enough information has accumulated in the fields of biology, chemistry, physics, and geology to justify serious speculation about life's origin.

Most of today's biochemists and geologists are convinced that life on earth began, a few billion years ago, with the appearance in earth's primeval seas of one or more carbon-containing molecules of something resembling nucleic acid, perhaps combined with something resembling protein, and capable of self-replication. The appearance of such a molecule (or molecules) does not require, these scientists believe, the intervention of supernatural power. It can be explained satisfactorily in terms of physical laws, combined with the laws of mathematical probability.

Such a view is deeply disturbing to a certain type of religious believer. In the United States there are still millions of Protestant fundamentalists, most of them in the South, who do not believe in evolution. These fundamentalists are convinced that about six thousand years ago, in a series of stupendous magic tricks, God created all living things. Millions of other devout Christians, Catholic and Protestant, accept the theory of evolution but believe that, at some moment in earth's history, several billion years ago, a special creative act of God caused the first living molecule (or molecules) to appear on earth.

Let me confess at once that I find something profoundly impious, almost blasphemous, about setting limits of any sort on the power of God to bring things about in any manner He chooses. If God creates a world of particles and waves, dancing in obedience to mathematical and physical laws, who are we to say that He cannot make use of those laws to cover the surface of a small planet with living creatures? A god whose creation is so imperfect that he must be continually adjusting it to make it work properly seems to me a god of relatively low order, hardly worthy of worship. The belief in a miraculous creation, miraculous in the sense that natural laws are momentarily suspended by a special act of God, is what I like to call the "superstition of the finger"—the belief that God periodically reaches into His universe, so to speak, to tinker around with it in various ways. It was precisely this superstition that made it so difficult for Christians of the nineteenth century to accept evolution. But as the scientific evidence for evolution became overwhelming, it finally dawned on most theologians that there was no reason whatever why they could not accept it; evolution was simply God's way of creating new forms of life.

Today it is hard to find a single biochemist or geologist, even among the most devoutly religious, who has the slightest doubt about the essential soundness of the theory of evolution. There may be many disagreements over details, but none over the broad outline. When a living organism makes a copy of itself, the copy is almost always, but not always, perfect. On rare occasions some type of radiation (such as ultraviolet light from the sun, cosmic rays, or radiation from radioactive substances in the earth) hits the nucleic acid helix and knocks its atoms into a slightly different pattern. The genetic code is altered; a copy is made that differs in some small, random way from the original. Usually the change is harmful to the organism. In that case the mutant and its offspring are less likely to survive and perpetuate the harmful change. When the change is beneficial, the mutant and its offspring have better-than-average survival chances. In this way "natural selection" causes slow modifications to take place over long periods of time. New "species" arise. Evolution is simply the process by which chance (the random mutations) cooperates with natural law to create living forms better and better adapted to survive.

If this union of nature and chance can be God's method of creating new species, why cannot a similar union of nature and chance be God's method of creating the first "living" molecules? Such a view does not make life any less wonderful or mysterious. As Loren Eiseley has said with such eloquence (at the end of his book *The Immense Journey*), it only makes the elementary particles more wonderful and mysterious. "If 'dead' matter has reared up this curious landscape of fiddling crickets, song sparrows, and wondering men," he writes, "it must be plain even to the most devoted materialist that the matter of which he speaks contains amazing, if not dreadful powers, and may not impossibly be, as Hardy has suggested, 'but one mask of many worn by the Great Face behind.' "

Let us travel back in our mind to those desolate, primordial ages, three or four billion years ago, when no living thing moved on the face of the earth or in its waters. How did the first "live" molecule come to be? Did God stretch out his hand and with his finger (I speak metaphorically) push together some atoms of carbon, oxygen, nitrogen, hydrogen, and sulfur into the pattern of a giant polymer capable of self-replication? We cannot say that such an event did not take place. But we *can* look for an explanation more dignified, more in keeping with a larger concept of deity.

Perhaps spores of living molecules, from somewhere else in the universe, fell into earth's oceans and found there an environment capable of supporting them. A number of scientists have favored such a theory. Svante Arrhenius, a famous Swedish chemist, wrote an entire book in defense of this view, *Worlds in the Making* (an English translation of which was published in 1908). In this book he argued that life on earth might have arisen from deep-frozen spores that had been propelled here through interstellar space by the pressure of radiation from the stars.

A similar idea, that living spores were carried to the earth by meteorites, has recently been revived by several studies of the composition of certain types of meteorites rich in carbon. In 1961 a group of American scientists reported they had found a number of complex hydrocarbons, very much like those found on earth in living things, in a sample taken from a meteorite owned by the American Museum of Natural History. Later that year another group of U.S. scientists found in meteorites some microscopic particles that may be fossils of simple plant life. One scientist announced that he had extracted *living* microorganisms from a meteorite, but the consensus among experts is that what he found were contaminants picked up from the earth's atmosphere. Biochemists are prepared to admit that meteorites may contain *fossil* evidence of once-living things. They are inclined to doubt strongly that life itself could survive the radiation hazards of a journey through space, either on a meteorite or in the form of free spores.

There is, however, no longer any doubt that fairly complex carbon compounds, so essential to life as we know it, have been formed by chemical processes outside the earth. On the morning of September 28, 1969, a meteor exploded over the town of Murchison in Australia. It was of a type called carbonaceous chondrite, extremely rich in carbon compounds. A team of scientists, headed by Cyril Ponnamperuma, a Sri Lankan biochemist, later found a variety of amino acids in a fragment of this meteorite. Since then, other meteorites have been found to contain amino acids.

In 1978 a carbonaceous chondrite meteor, found on top of Antarctic ice, was shown by NASA scientists to contain methane. It was the first proof that methane exists outside our solar system.

Amino acids had been reported before 1969 in meteorites, but the prevailing opinion was that they were the result of contamination. "You have only to make a thumbprint on a beaker and shake with water to obtain amino acids," was how Ponnamperuma put it. But in the case of the Murchison meteorite this possibility was ruled out. The main reason for ruling it out was that each amino cid showed most equal amounts of left- and right-handed forms. Had they been of earthly origin, all of them would have been left-handed.

At about the same time that amino acids were found in meteorites, another startling discovery of a similar nature was made by radio astronomers. They obtained strong evidence that dozens of organic molecules were present in interstellar space. Billions of alcohol molecules, for example, are drifting about in the constellation of Sagittarius. Formaldehyde, hydrogen cyanide, and formic acid have also been identified. It seems as if there are forces capable of creating complex organic molecules almost anywhere in the universe.

On a more fanciful level, science-fiction writers have imagined higher forms of intelligence traveling about the cosmos and "seeding" planets that have physical and chemical conditions favorable to life. Thomas Gold, the English astrophysicist, once suggested that life on earth may have arisen from microbes in the garbage left by nonterrestrial astronauts who visited our planet several billion years ago.

Most biochemists today reject the view that life on earth had an extraterrestrial origin. Their reasons are not so much the lack of evidence for this view, or the difficulties of explaining how life could withstand cosmic radiation on its trip through space; their views rest mainly on the growing evidence that living organisms could easily have arisen spontaneously right here on earth.

"Spontaneous generation," in the sense of a constantly occurring production of living things from nonliving matter, was vigorously defended by many great biologists from the time of Aristotle up to the time of Pasteur. Before the theory of evolution became well established, it was widely thought that all sorts of living forms, even mice, were generated spontaneously from ooze and slime or decaying animal tissue. In Pasteur's day most chemists believed that microbes were spontaneously generated in stagnant water. In a series of simple but brilliantly conceived experiments Pasteur proved once and for all that this was not the case. Biologists who thought they had found evidence for it had simply not been careful enough to prevent airborne microbes from sneaking into their flasks. Today no reputable biochemist thinks that microorganisms are being generated, anywhere on earth, from nonliving matter. The most that could happen would be the occasional appearance of primitive half-living molecules on the sea's surface, where they would be quickly gobbled up by living microorganisms. Even this seems extremely unlikely.

Nevertheless, biochemists believe that spontaneous generation must have taken place at least once, 3 or 4 billion years ago, when chemical and physical conditions on the earth were vastly different from what they are now. The saltless oceans probably contained great quantities of ammonia and carbon dioxide. No free oxygen was then present in the atmosphere to form a protective layer of ozone that would shield the earth from the powerful ultraviolet radiation of the sun. This radiation, beating down on primeval waters, could have supplied enough energy to change some of the simple hydrocarbon molecules in the sea to more complex chain molecules. Other sources of energy could have been the earth's heat, which may have been much greater than now, the lightning that must have played over the sea's surface, radiation from radioactive substances within the earth, and radiation from cosmic rays. Over a long period, perhaps more than a billion years, with the vast oceans swirling and churning, it is not unreasonable to suppose that millions of different complex carbon-containing molecules could have taken shape.

Science writers (and some scientists) have a compulsion to overdramatize the sudden appearance of one molecule, perhaps a molecule of nucleic acid, capable of self-replication: a kind of chemical Adam that started the drama of evolution. No one can say that is not the way it happened, but more likely there was no such dramatic turning point. Self-replication is a matter of degree. Organic molecules capable of partial, incomplete replication may have appeared first, then rapidly multiplied by forming millions of crude copies. Even if we knew the details of what happened throughout these millions of years that preceded the beginning of the fossil record, we might be unable to point to any one year, or even to a period of a thousand years, and say, "That is where life began." There may have been a gradual increase in complexity, along a continuum, until finally organic molecules began to appear with a structure similar to that of the nucleic-acid molecules found today in living things.

Many scientists, with strong emotional ties to the superstition of the finger, have scoffed at the notion that fortuitous combinations of organic molecules in primordial seas could have produced a combination as structurally complex as nucleic acid. One of the earliest, most eloquent of such scoffers was Francis Robert Japp, a nineteenth-century Scottish chemist at the University of Aberdeen. In a widely discussed address, "Stereochemistry and Vitalism" (printed in *Nature*, September 8, 1898, pages 452*ff*.), he gave an excellent summary of Pasteur's work on stereoisomers, then launched into a vigorous defense of the supernatural origin of the first asymmetric molecules. Stereoisomerism of a single handedness could not possibly have arisen, he maintained, from the blind operations of the symmetric forces of nature.

"Only the living organism with its asymmetric tissues," Japp declared, "or the asymmetric products of the living organism, or the living intelligence with its conception of asymmetry, can produce the result. Only asymmetry begets asymmetry. . . . If these conclusions are correct, as I believe they are, then the *absolute origin* of the compounds of one-sided asymmetry to be found in the living world is a mystery as profound as the absolute origin of life itself. . . . No fortuitous concourse of atoms, even with all eternity for them to clash and combine in, could compass this feat of the formation of the first optically active organic compound. Coincidence is excluded, and every purely mechanical explanation of the phenomenon must necessarily fail."

Japp's lecture sparked considerable controversy among readers of *Nature*. Many distinguished scientists and thinkers, including Herbert Spencer, Karl Pearson, and George FitzGerald (the man who worked out the mathematics of the Lorentz-FitzGerald contraction theory in relativity), wrote letters of protest that were printed in *Nature* together with numerous rebuttals by Japp.[1]

Professor Japp's arguments were revived in Lecomte du Noüy's widely read book *Human Destiny* (1947) and repeated in his later book *The Road to Reason* (1949). The odds against the chance formation of a complex, asymmetric, organic molecule are so great, maintains du Noüy, as to amount to virtual certainty that the event could not have taken place without divine intervention. One would as soon expect a Shakespearean play to be typed out by those monkeys banging on typewriter keys. The English astronomer Arthur Stanley Eddington said it this way:

> There once was a brainy baboon
> Who always breathed down a bassoon,
> For he said, "It appears
> That in billions of years
> I shall certainly hit on a tune."[2]

The probability of Eddington's baboon hitting on a tune is difficult to estimate without first defining what is meant by the word *tune*. One would not expect a chimpanzee, dribbling or smearing paint on canvas, to produce a replica of the "Mona Lisa," but if the word *painting* includes all the products of contemporary abstract expressionists, then it is difficult for a chimpanzee, properly instructed, *not* to produce a painting. A similar semantic difficulty is encountered in trying to estimate the probability of the fortuitous appearance of a complex organic molecule. How complex is "complex"?

In 1952 an American chemist named Stanley L. Miller, who was just twenty-three at the time, actually produced some fairly complex amino acids by a simple technique designed to test a theory suggested by his teacher, the noted chemist Harold Urey. In a flask he put a mixture of water, ammonia, methane, and hydrogen—a mixture which Professor Urey believes to be similar to the mixture of elements in the earth's primordial oceans and atmosphere. Energy was supplied by an electric discharge passed through the mixture continually for one week. When the mixture was analyzed at the end of that time, Miller found in it various organic compounds, including amino acids, which had not been there before.

It is true that this is a long way from producing nucleic acid or even protein, but amino acids are the asymmetric building blocks of protein. On the basis of Japp's and du Noüy's methods of computing odds, even a humble amino acid should not have been expected in such an incredibly short period as one week, and in such a picayune amount of chemicals. The experiment was a milestone in the history of theories about the origin of life.[3] It has since been repeated by many other scientists, using slightly different mixtures and energy sources.

In 1963 Ponnamperuma and his associates succeeded in similarly producing one of the chief components of nucleic acid. A beam of high-energy electrons was shot through a mixture of hydrogen, ammonia, methane, and water vapor for about forty-five minutes. In the mixture were found minute amounts of adenine, one of the five nucleotide bases. More recently, Sidney W. Fox and Kaoru Harada, at Florida State University, were able to synthesize thirteen different kinds of amino acids by using nothing but heat (about 1,000 degrees centigrade) as their energy source. In 1967 Arthur Kornberg and his colleagues at Stanford University artificially produced the active, infectious DNA inner core of a virus. When it was injected into living cells, the infected cells began producing viruses indistinguishable from natural ones. In 1969 two American research teams independently synthesized ribonuclease, an enzyme made of nineteen types of amino acids. After these experiments not a single scientist has dared to argue that complex organic compounds could not result from the operation of chance and natural laws.

Where did Japp and du Noüy go wrong? The main loophole in their argument is this: much more than chance was operating on those swirling compounds in earth's primordial seas—there were also natural laws of physics and chemistry. Spill a bag of beans on a table top and it is unlikely they will form a pattern with regular hexagonal symmetry. But we know that when water freezes during a snowstorm the molecules form such hexagonal patterns by the millions. The reason is, of course, that electrical forces of attraction and repulsion are operating between the molecules in such a way as to make such striking patterns not only possible but extremely probable.

Isaac Asimov has put it this way: Suppose we take the atoms of hydrogen and oxygen and combine them *at random* to form molecules with three atoms each, assuming that they can form any combination, such as HHH, HHO, HOH, HOO, and so on. From this mixture we extract ten molecules at random. What are the chances that all ten will be HHO; that is, molecules of water? As Asimov works it out, the chances are about 1 in 60,000,000. We all know, however, that if we actually performed such an experiment the atoms would *not* combine at random. *All* the molecules would be molecules of water, because that is the only three-atom combination chemically possible for hydrogen and oxygen atoms. What Japp and du Noüy failed to take into account was the operation of natural laws. Atoms, as Asimov says, are not like sticky marbles which, when shaken in a barrel, can stick together any old way. They combine only in a manner determined by physical laws.

The plain fact is that we do not know enough about the electrical forces operating on the atoms in a soup of carbon compounds, under conditions that prevailed on earth before evolution got under way, to make any meaningful estimate of the probability of a particular combination. Certain combinations may be impossible, others extremely probable. Du Noüy's greatest mistake was in trying to estimate the probability of a self-replicating molecule on the assumption that atoms combined by blind chance. He should have asked himself, writes Asimov, what the chances would be that such a molecule could be built up by the "*unblind* workings of chance"—that is, the workings of chance in concert with the laws of chemistry and physics. For all we know, primordial conditions may have made it difficult for amino acids *not* to form, and once formed, difficult *not* to join into complex chains.

We do know that only a week was required for the random production of asymmetric amino acids in the small amount of chemicals in Miller's flask. Given a billion years of time, a chemical mixture as large as the earth's seas and atmosphere, and various energy sources more intense than today, who can say that no self-replicating molecules could have formed fortuitously? For all we know, they may have formed by the billions. Perhaps first the amino acids came together to form billions of different protein molecules. Perhaps a molecule of nucleic acid, or something resembling it, then latched onto a bit of protein and something came into being that was capable of copying itself with fair accuracy whenever it found the proper proteins. In a few thousand or a few million years (all of this is sheer guesswork) the primordial soup may have swarmed with these primitive, half-living organisms. The great epic of evolution would then have been under way.

Notes

1. These letters are still worth reading and by no means hopelessly out of date. Letters by the following writers, in issues of *Nature* in 1898, are of special interest: Giorgio Errera, October 27, p. 616; Karl Pearson, September 22, p. 495; George F. FitzGerald, Clement O. Bartrum, October 6, p. 545; Herbert Spencer, October 20, p. 592; Herbert Spencer, Karl Pearson, Percy F. Frankland, November 10, p. 29. Additional letters appear on November 17, p. 53. Japp's final rebuttal is December 1, p. 101.

2. From Eddington's *New Pathways in Science* (Cambridge, 1935), chapter 3. He gives no source for the limerick and is suspected of having written it himself.

3. For Miller's own account of his historic experiment see "A Production of Amino-acids under Possible Primitive Earth Conditions," *Science*, Vol. 117, 1953, pp. 528*ff*.

16. THE ORIGIN OF ASYMMETRY

In a way it is amusing to find so many well-meaning theists cringing with horror these days at theories designed to bridge the gap between nonlife and life by the operation of "unblind chance"—the union of chance and natural law. It is amusing because it is easier to imagine *this* gap bridged than many of the later gaps in the history of life on earth. For example, chlorophyll had to be discovered, as the means by which living units (plants) could use solar energy to manufacture starches and fats. Single-cell animals had to discover the shortcut of eating the plants. Death and sex had to be invented by many-celled organisms capable of growing old and ceasing to function as a cooperative colony of cells. Animals had to discover how to eat other animals. Above all, an intelligent species of animal had to evolve—a species so brainy that it has discovered a way of blowing up the earth and bringing the entire evolutionary process to an end. To an extra-terrestrial observer some of these steps might well seem less probable than the initial step from lifeless to living matter.

How excited and pleased Pasteur would have been if he could have known of Miller's famous experiment! Though himself a theist, Pasteur was convinced that God created life on the earth by just such a combination of chemicals, forces, and chance. He recognized also, as we have seen, that the organic compounds of living things are optically active—that is, they possess an internal asymmetry capable of twisting planes of polarized light. He was impressed, as well he should have been, with the fact that outside of living tissues asymmetric compounds are always found in racemic form: a mixture of right and left molecules. Only in living tissues do organic compounds have a pure handedness.

Pasteur believed that, if he could only discover how nature introduced this asymmetry into organic compounds, he would be close to the secret of life itself. It seemed to him probable that some sort of asymmetry in the earth's environment provided asymmetrical forces which must have acted on the first living units and given them an asymmetrical twist. "Life, as manifested to us," he wrote, "is a function of the asymmetry of the universe and of the consequences of this fact. . . . I can even imagine that all living species are primordially, in their structure, in their external forms, functions of cosmic asymmetry."

Pasteur believed that magnetism exhibited a glaring instance of natural asymmetry in the universe. If you place a magnetic needle above a wire through which a current is flowing directly away from you, the needle will assume a position at right angles to the wire. Instead of pointing with its north pole to the right as often as it points to the left, the needle always points to the left. As we shall learn in chapter 19, this only *seems* to be an asymmetric phenomenon; but in Pasteur's time magnetism was poorly understood and all the scientists of his day thought that magnetism possessed a fundamental asymmetry, in contrast to symmetric forces such as gravity and inertia. Acting on this belief, Pasteur performed a variety of fantastic experiments. For example, he grew crystals between the poles of powerful magnets, hoping this would cause a majority of crystals to form with a certain handedness. He was disappointed in his complete failure to induce asymmetry, either in crystals or compounds, by the application of magnetism.

Another possibility that occurred to Pasteur was that the passage of the sun through the sky from east to west might exert an asymmetric influence on substances. Since the earth has north and south magnetic poles, perhaps the sun's movement, combined with terrestrial magnetism, would induce asymmetry. By using clever arrangements of mirrors and clockwork mechanisms, he was able to grow plants under conditions in which sunlight actually passed over the plant from west to east instead of the usual direction. Pasteur hoped this would cause the plant to grow optically active substances that would rotate polarized light in a direction opposite to what would normally be expected. Again, the results were disappointingly negative.

To this day no one knows how the first half-living molecule, or the first half-living molecules, got their particular handedness. As we have seen, all amino acids in living tissues have the same left-handed twist. This is sufficient to account for the uniformity in the handedness of all protein helices. The same is true of the nucleotides that impart their left-handed twists to coils of nucleic acid. If the first molecule capable of self-replication happened, by sheer accident, to be left-handed rather than right-handed, then of course all its copies would be left-handed. This could explain the universality of left-handedness in the amino acids and nucleotides. Asymmetry begets asymmetry. The "Adam" molecule would link itself only to proteins of the same handedness; then copies would pass that handedness on down to all later copies. Had the first self-replicating molecule been of a different handedness, all life would have "gone the other way."

It is also possible that millions of primitive, half-living, partly self-replicating molecules arose in the earth's primordial "hot soup," and that some asymmetric feature of the environment gave all of them, or a majority, a left-handed twist. Since Pasteur's day many theories along such lines have been devised. It has been suggested that life began in one hemisphere, where Coriolis forces in some way provided the required twist. Had life started in the other hemisphere, according to this theory, amino acids would now be right-handed instead of left. This theory has not won wide acceptance.

A better suggestion: elliptically polarized light (a type of polarization that results when light is reflected from surfaces) may have combined with the earth's magnetic field to supply the twist. Laboratory experiments with elliptically polarized light in magnetic fields have been successful in synthesizing one-handed compounds. Light reflected from the earth's primeval oceans could have had this sort of polarization, but most biochemists do not believe the effect would have been strong enough to give a significant left-handed bias to the earth's primitive organic molecules.

In 1931 a Russian scientist named V. Vernadski made a startling suggestion. Some astronomers believe that the moon was once part of the earth. At the time the moon was separated from the earth, Vernadski reasoned, perhaps a colossal wrench of some sort, asymmetrical in nature, imparted a left-handed twist to organic molecules then being formed.

Still another suggestion was advanced by physicist Joseph Rush in his splendid book *The Dawn of Life* (Signet, 1962). Perhaps self-replicating molecules of both types of handedness evolved in the primordial broth. Each molecule could feed only on molecules of its own handedness. Then a mutation of one left-handed molecule gave it the ability to feed on *both* left- and right-handed compounds, possibly even on its right-handed living competitors. As it multiplied, its descendants would have a strong competitive advantage over rivals that could feed only on their own type of handedness. Eventually only the more versatile mutant species would remain, and of course it would pass on its left-handedness to all its progeny.

Even without such a mutation it is possible that molecules of a certain handedness might outbreed their mirror images. If you toss a penny a hundred times, it is extremely improbable that you will get *exactly* fifty heads and fifty tails. It is similarly improbable, if asymmetric compounds were formed in large numbers, that the number of right-handed ones would exactly balance the number of left-handed ones. Whichever handedness predominated might gain a competitive advantage simply by virtue of its larger numbers. For example, a sudden change in the environment might result in widespread destruction of both types, with the more numerous type having a better chance to survive.

All these theories are highly tentative. No one can claim to know how life on earth acquired its particular set of asymmetries. Whatever happened a few billion years ago, most biologists are convinced it is no longer occurring. For one thing, as previously noted, newly created, half-living molecules on the surface of the sea would quickly be devoured by microorganisms. For another thing, conditions on the earth are not at all the same as they were in early geologic ages. Plants have filled the atmosphere with oxygen. This screens off much of the powerful ultraviolet radiation of the sun—radiation which may have been essential as an energy source for the formation of the first organic molecular chains. Whatever happened probably ceased to happen several billion years ago.

Pasteur favored the view that some sort of basic asymmetry in the earth's environment, possibly one that still prevailed, was responsible for the original left–right bias of organic molecules. He was groping as best he could in an area of vast darkness. The sharp conflict between the symmetry of nonlife and the asymmetry of life fascinated him. He had a strong intuitive hunch that in some yet unknown fashion there was a fundamental asymmetry at the very heart of the universe itself. *"L'univers,"* he wrote, *"est dissymétrique."* He was wrong in thinking that magnetism reflected this universal cosmic handedness. Nevertheless, we shall see in a later chapter that Pasteur's hunch may yet turn out to be true—true in a way which in his day could not be conceived. But first we must pause for a brief philosophical look at asymmetry and the fourth dimension and at a perplexing problem in communication that will make it easier to understand the physics involved in the book's remaining chapters.

17. THE FOURTH DIMENSION

Immanuel Kant, the great German philosopher of the eighteenth century, was the first eminent thinker to find a deep philosophical significance in mirror imagery. That an asymmetric object could exist in either of two mirror-image forms seemed to Kant both puzzling and mysterious. Before discussing some of the implications Kant drew from left–right asymmetry, let us first see if we can recapture something of the mood in which he approached this topic.

Imagine that you have before you, on a table, solid models of the enantiomorphic polyhedrons shown in Figure 42. The two models are *exactly alike* in all geometrical properties. Every edge of one figure has a corresponding edge of the same length on the other figure. Every angle of one figure is matched by a duplicate angle on the other. No amount of measurement or inspection of either figure will disclose a single geometrical feature not possessed by the other. They are, in a sense, identical, congruent figures. Yet clearly they are *not* identical!

This is how Kant expressed it, in section 13 of his famous *Prolegomena to All Future Metaphysics*: "What can more resemble my hand or my ear, and be in all points more like, than its image in the looking-glass? And yet I cannot put such a hand as I see in the glass in the place of its original. . . . "

That two objects can be exactly alike in all properties, yet unmistakably different, is certainly one reason why the looking-glass world has such an eerie quality for children and for primitive people when they encounter it for the first time. Of course, the major source of spookiness is simply the appearance behind the glass of a world that looks as real as the world in front yet is completely illusory. If you want to puzzle and fascinate a small child, stand him in front of a large wall mirror at night, in a dark room, and hand him a flashlight. When he shines the flashlight into the mirror the beam goes straight into the room behind the glass and illuminates any object toward which he aims it. This strong illusion of a duplicate room is spooky enough, but it grows even spookier when one becomes aware of the fact that everything in the duplicate room "goes the other way." It is the *same* room, yet it isn't.

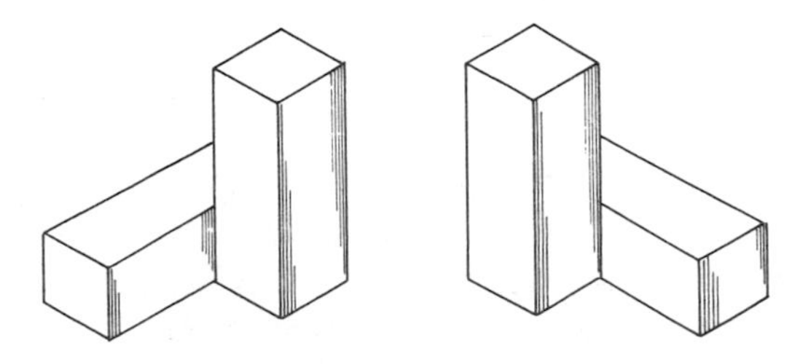

Figure 42. Enantiomorphic polyhedrons.

Exactly what Kant made of all this is a tangled, technical, controversial story. During the past few decades Kant has been so mercilessly pilloried by Bertrand Russell and other leading philosophers of science that readers on the sidelines are apt to think of Kant as a woolly-brained metaphysician who had little comprehension of mathematics and science. The fact is that Kant was well trained in the science and mathematics of his day. He began his career as a lecturer on physics, and most of his early writings were on scientific topics. Like Alfred North Whitehead, he turned from mathematics and physics to the construction of a metaphysical system only in his later years. Whatever one may think of his final conclusions, there is no denying the importance of his ground-breaking contributions to the philosophy of modern science.

Kant's first published paper, "Thoughts on the True Estimation of Living Forces" (1747), contains a remarkable anticipation of *n*-dimensional geometry. Why, he asks, is our space three-dimensional? He concludes that somehow this is bound up with the fact that forces such as gravity move through space, from a point of origin, like expanding spheres. Their strength varies inversely with the square of the distance. Had God chosen to create a world in which forces varied inversely with the *cube* of the distance, a space of four dimensions would have been required. (Similarly, though Kant did not mention it, forces in 2-space, moving out from a point source in expanding circles, would vary only inversely with the distance.) Kant here adopted a view of space that had been put forth a century earlier by Gottfried Wilhelm von Leibnitz, the great German philosopher and mathematician. Space has no reality apart from material things; it is nothing more than an abstract, mathematical description of relations that hold between objects. Although the notion of a fourth dimension had occurred to mathematicians, it had been quickly dropped as a fanciful speculation of no possible value. No one had hit on the fact that an asymmetric solid object could (in theory) be reversed by rotating it through a higher space; it was not until 1827, eighty years after Kant's paper, that this was first pointed out by August Ferdinand Moebius, the German astronomer for whom the Moebius strip is named. It is surprising, therefore, to find Kant writing as early as 1747: "A science of all these possible kinds of space [spaces of more than three dimensions] would undoubtedly be the highest enterprise which a finite understanding could undertake in the field of geometry." He adds, "If it is possible that there are extensions with other dimensions, it is also very probable that God has somewhere brought them into being; for His works have all the magnitude and manifoldness of which they are capable." Such higher spaces would, however, "not belong to our world, but must form separate worlds."

In 1768, in a paper "On the First Ground of the Distinction of Regions in Space," Kant abandoned the Leibnitzian view of space for the Newtonian view. Space is a fixed, absolute thing—the "ether" of the nineteenth century—with a reality of its own, independent of material objects. To establish the existence of such a space, Kant turned his attention toward what he called "incongruent counterparts"—asymmetric solid figures of identical size and shape but opposite handedness, such as snail shells, twining plants, hair whorls, right and left hands. The existence of such twin objects, he argued, implies a Newtonian space. To prove it, he made use of a striking thought experiment, which can be stated as follows.

Imagine that the cosmos is completely empty save for one single human hand. Is it a left, or a right hand? Since there are no intrinsic, measurable differences between enantiomorphic objects, we have no basis for calling the hand left or right. Of course, if you imagine yourself looking at the hand, naturally you will see it as either left or right, but that is equivalent to putting yourself (with your sense of handedness) into 3-space. You must imagine the hand in space to be completely removed from all relationships with other geometrical structures. Clearly, it would be as meaningless to say that the hand is left or right as it would be to say it is large or small, or oriented with its fingers pointing up or down.

Suppose now that a human body materializes in space near the hand. The body is complete except for both hands; they have been severed at the wrist and are missing. It is evident that the hand will not fit both wrists. It will fit only one—say the left wrist. Therefore it is a left hand. Do you see the paradox confronting us? If it proves to be a left hand, by virtue of fitting the left wrist, it must have been a left hand *before* the body appeared. There must be some basis, some ground, for calling it "left" even when it is the sole object in the universe. Kant could see no way of providing such a ground except by assuming that space itself possessed some sort of absolute, objective structure—a kind of three-dimensional lattice that could furnish a means of defining the handedness of a solitary, asymmetric object.

A modern reader familiar with n-dimensional geometry should have little trouble seeing through the verbal confusion of Kant's thought experiment. In fact, Kant's error was effectively exposed by an episode in Johnny Hart's syndicated comic strip called *B.C.*, in newspapers of July 26, 1963. One of Hart's cavemen has just invented the drum. He strikes a log with a stick held in one hand and says, "That's a left flam." Then he hits the log with a stick in his other hand and says, "That's a right flam."

"How do you know which is which?" asks a spectator.

The drummer points to the back of one hand and replies, "I have a mole on my left hand."

Let us see how this relates to Kant's error. Imagine that Flatland contains nothing but a single, flat hand. It is true that it is asymmetrical, but it is meaningless to speak of it as left or right if there is no other asymmetric structure on the plane. This is evident from the fact that we in 3-space can view the hand from either side of the plane and see it in either of its two mirror-image forms. The situation changes if we introduce a handless Flatlander and *define* "left" as, say, the side on which his heart is located. This by no means entails that the hand was "left" or "right" before introducing the Flatlander, because *we can introduce him in either of two enantiomorphic ways.* Place him in the plane one way, the hand becomes a left hand. Turn him over, place him the other way, and the hand becomes a right hand—"right" because it will fit the wrist on the side opposite the heart.

Does this mean that the hand alters its handedness, or that the Flatlander's heart magically hops from one side of his body to the other? Not at all. Neither the hand nor the Flatlander changes in any respect. It is simply that their relations to each other in 2-space are changed. It is all a matter of words. "Left" and "right" are words which mean, as Humpty Dumpty said, whatever we want them to mean. The solitary hand can be labeled with either term. So can the sides of a solitary Flatlander. It is only when the two asymmetric objects are present in the same space, and a choice of labels has been made with respect to one, that labels applied to the other cease to be arbitrary.

It is the same in 3-space. Not until we introduce the handless body, with the understanding that "left" is the side the heart is on, do we have a basis for deciding what to call the hand. If the body is "turned over" by rotating it through 4-space, the hand's label automatically changes. Suppose we first label the solitary hand, calling it, say, a "right" hand. When the body appears, its "right" wrist will be, by simple definition, the wrist on which the hand fits. The important point is that the initial choice of terms is wholly arbitrary. Hart's caveman who chose to call one hand "left" because it had a mole on it was making a completely rational first step in defining handedness. The humor of the strip lies in the way the caveman phrased his reply. Instead of saying that he knew the difference between left and right flams because he had a mole on his left hand, he should have said: "Because I have decided to call 'left' the hand that has a mole on it." There is nothing paradoxical about such a situation, therefore no need to introduce Newton's absolute space.[1]

Actually, even a fixed, Newtonian ether is no help in providing a label for the solitary hand unless the structure of space itself is somehow asymmetrical. If the hand floats inside a spherical, cylindrical, or conical cosmos, or in an infinite space crisscrossed with the lines of a cubical lattice, we are no better off than before. If the cosmos has the shape of one enormous human hand, the situation changes. We could call the cosmic hand "right" (or "plus" or "Yin"); then, if the solitary human hand is of opposite handedness, we are forced to call it "left" (or "minus" or "Yang"). We could also define the hand's handedness on the basis of an asymmetric "grain" in space, a submicroscopic lattice of geodesics (straightest possible paths) like the asymmetric lattice of quartz or cinnabar. In later chapters we will see that such speculations are now of the highest interest in connection with recent discoveries about the asymmetric behavior of certain elementary particles.

Kant himself soon realized that his thought experiment proved nothing. In later, more mature reflections he combined the views of Newton and Leibnitz into a novel synthesis of his own, intimately bound up with his transcendental idealism. Newton was right, he argued, in regarding space as independent of material bodies, but Leibnitz was also right in denying reality to space. Space is independent of bodies precisely because it is not real; it is ideal, subjective, a mode by which we view a transcendent reality utterly beyond our comprehension.

Space and time are like the two lenses in a pair of glasses. Without the glasses we could see nothing. The actual world, the world external to our minds, is not directly perceivable; we see only what is transmitted to us by our space-time spectacles. The *real* object, what Kant called the *Ding-an-sich* or Thing-in-itself, is transcendent, beyond our space-time, completely unknowable. ("The solution of the riddle of life in space and time lies outside space and time," writes Ludwig Wittgenstein in his *Tractatus Logico-Philosophicus,* 6.4312.) We experience only our sensory perceptions: what we see, hear, feel, smell, taste. These perceptions are, in a sense, illusions. They are shaped and colored by our subjective sense of space and time, as the color of an object is influenced by colored glasses or the shape of a shadow is influenced by the surface on which it falls.

> Space is a swarming in the eyes; and time,
> A singing in the ears.[2]

"What then is the solution?" Kant asks in his *Prolegomena.* "These [mirror-image] objects are not presentations of things as they are in themselves, and as the pure understanding would cognize them, but they are sensuous intuitions, *i.e.,* phenomena, the possibility of which rests on the relations of certain unknown *things in themselves* to something else, namely, our sensations."

In trying to get at the meaning of statements made by philosophers who lived many generations ago, it is sometimes worth the risk to try to rephrase the statements in current terminology and in the light of current knowledge. Of course, it is highly speculative. Nevertheless, I think that if Kant were alive today he would make his point somewhat as follows.

Eighteenth-century mathematicians, as we have seen, had not yet discovered that Euclidian geometry could be extended to any number of dimensions. A straight line, one foot in length, is a one-dimensional figure. In two dimensions the corresponding figure is a square, one foot on a side. In three dimensions it is a cube, one foot on a side. This can be generalized by adding as many new dimensions as one wishes. A hypercube is a cube, one foot on a side, which extends in four directions, each direction at right angles to the other three. The mathematician can work out the geometrical properties of such a cube. There is no reason why a four-dimensional world could not exist, containing material hypercubes, or for that matter a world of five dimensions or six or seven. The hierarchy is endless. At each level the geometry is Euclidian—as valid and consistent as the familiar plane and solid Euclidian geometry taught in high school.

Mathematical techniques can uncover the properties of figures in these higher Euclidian spaces, but our minds are firmly trapped in a Euclidian 3-space, which is united with the single onrushing arrow of time. We find it impossible to conceive of a thing existing without extension in three spatial dimensions and duration in the one dimension of time. Perhaps with the right sort of training, or in some future age when the mind of man has evolved into a more powerful tool, one might learn to think in four spatial dimensions. At present we cannot do so. We see the world through our space-time spectacles: one lens is one-dimensional time, the other is three-dimensional space. We cannot visualize in our brain the structure of a hypercube or any other 4-space structure. We can only visualize 3-space structures that endure—which move along the single track of time.

Suppose, however, that there is a transcendent world, a world of 4-space, inaccessible to our senses and beyond our powers to imagine. How would a hyperperson, in such a hyperworld, view two solid asymmetric objects such as the polyhedrons in Figure 42 that are mirror images of each other? The mathematician can give a clear and unambiguous answer: The polyhedrons would appear identical, each superposable on the other.

To understand this, imagine yourself looking down on a world of 2-space and seeing the two asymmetric shapes shown in Figure 43. Flatlanders living on the plane would be just as puzzled by those two

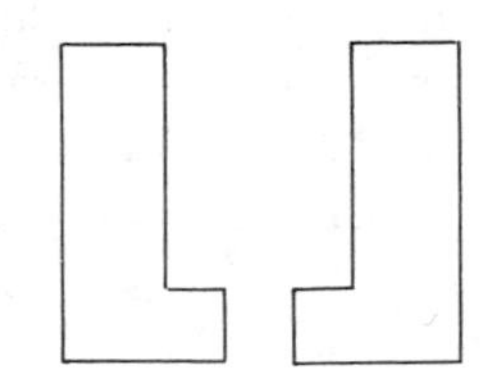

Figure 43. Enantiomorphic polygons.

figures as Kant was puzzled by his ears and their mirror reflections. How can two figures be so alike, the Flatlanders ask themselves, and yet be nonsuperposable? We who live in 3-space can understand. They *are* alike. It is only because the poor Flatlanders are trapped in 2-space, seeing things only through their 2-space Euclidian lenses, that they cannot see that the two shapes *are* superposable. We can prove that they are simply by picking one up, turning it over, and fitting it, point for point, on the other. If we return the reversed figure to the plane, next to the other one, the two figures will be seen by the Flatlanders as identical in every respect, including their handedness. Since the Flatlanders cannot conceive of 3-space, they will think a miracle has occurred. A rigid, asymmetric object has been changed to its mirror image! Yet we have done nothing to the object. We have not stretched, damaged, or altered it in any way. We have only altered its orientation in 2-space—its position relative to other objects in that space.

The two asymmetric polyhedrons in Figure 42 are similarly identical and superposable. It is only because we cannot see them through the transcendent spectacles of 4-space that we think they are not alike. If we could rotate one of them through hyperspace—turn it over, so to speak, through a fourth dimension—we would have a pair of congruent polyhedrons of the same handedness.

Kant did not, of course, express such views. Nevertheless, I think that if one makes a serious, well-informed attempt to put himself into the center of Kant's final vision of existence, he will find it not frivolous to suppose that Kant might have argued in this way had the mathematical knowledge of the twentieth century been available to him.

Leibnitz also had, I am persuaded, an intuitive grasp of the then-as-yet-undiscovered higher Euclidian spaces. He once considered the question of what would happen if the entire universe were suddenly reversed so that everything in it became its mirror image. He concluded that nothing would happen. It would be meaningless to say such a reversal had occurred, because there would be no way one could detect such a change. To ask why God created the world this way and not the other is to ask, Leibnitz said, "a quite inadmissible question."

When we view this question in the light of the various levels of Euclidian space, we see at once that Leibnitz is right. To "reverse" an entire Flatland on a sheet of paper, all we need do is turn the paper over and view the figures from the other side. We do not even have to turn the paper. Imagine a Flatland on a vertical sheet of glass standing in the center of a room. It is, say, a left-handed world when you view it from one side of the glass. Walk around the glass, you see it as a right-handed world.

EXERCISE 11: *When Mrs. Smith started to push open the glass door at the entrance to the bank, she was puzzled to see the word TUO printed on the door in large black letters. What does the word mean?*

Flatland itself does not change in any way when you view it from another side. The only change is in the spatial relation, in 3-space, of Flatland and you. In precisely the same way, an inhabitant of 4-space could view one of our kitchen corkscrews from one side and see a right-handed helix, then change his position and see the same corkscrew from the other side as a left-handed helix. If he could pick up one of our corkscrews, turn it over, and replace it in our continuum, it would seem to us a miracle. We would see the corkscrew vanish, then reappear in reflected form.

Enantiomorphic objects are identical not only in all metric properties; they are also topologically identical. Even though a right-handed knot in a closed loop cannot be deformed into a left-handed one, the two are topologically equivalent. Very young children seem to grasp this more readily than adults. Jean Piaget and Bärbel Inhelder, in their book *The Child's Conception of Space* (Humanities Press, 1956) report on strong experimental evidence that children actually recognize topological properties before they learn to recognize Euclidian properties of shape, including the distinction between left and right forms. When asked to copy a triangle, for example, very young children often draw a circle. The angles and sides of the triangle are less noticeable to them than the property of being a closed curve. They will see no difference between colors that go in a certain order clockwise around a circle and a circle on which the same colors go counterclockwise in the same order. Their untrained minds seem to sense that the two circles are identical: not that they realize that one can be turned over to become like the other, but rather that they see no difference to begin with. This may explain why even strongly right-handed children so often print letters backward, or sometimes entire words.

Perhaps our minds are potentially more flexible than Kant suspected. Our inability to visualize 4-space structures such as the hypercube may be due solely to the fact that all our memories are derived from experiences in a 3-space world. With suitable training toys, could a child learn to think in 4-space pictures? The question has been discussed seriously by a number of mathematicians and of course it is a familiar science-fiction gimmick, notably in Lewis Padgett's much anthologized tale "Mimsy Were the Borogoves."

Are there mirror-image forms among the hypersolids of 4-space—that is, shapes identical in all respects except their handedness? Yes, this duality exists on every level. In one dimension, figures are mirrored by a point; in two dimensions, by a line; in three dimensions, by a plane; in four dimensions, by a solid. And so on for the higher spaces. In every space of n dimensions the "mirror" is a "surface" of $n - 1$ dimensions. In every space of n dimensions an asymmetric figure can be made to coincide with its reflection by rotating it through a space of $n + 1$ dimensions. Perhaps our imaginary twentieth-century Kant would put it this way: Only the "pure understanding" of God Himself, who stands outside space and time, would see all pairs of enantiomorphic structures, in all spaces, as identical and superposable.

H. G. Wells was the first to base a science-fiction story on the reversal of an asymmetric solid structure by turning it around in 4-space. In "The Plattner Story," one of Wells's best, a young chemistry instructor named Gottfried Plattner explodes a mysterious green powder that blows him straight into 4-space. What he sees during the nine days that he lives in the dark "Other World," with its huge green sun and unearthly inhabitants, you will have to discover for yourself by reading Wells's story. It can be found in a collection, *28 Science Fiction Stories*, by H. G. Wells (Dover, 1952). After nine days in 4-space, Plattner slips on a boulder, the bottle of green powder explodes in his pocket, and he is blown back into 3-space. But his body has been turned over. His heart is now on the right. He writes a mirror script with his left hand.[3]

The drifting, mute figures in Wells's 4-space are the souls of those who once lived on earth. This notion that departed souls inhabit a higher space was a common one in the spiritualist circles of Wells's day; from time to time mediums actually were asked to change an asymmetric object to its mirror image as proof they were in genuine contact with 4-space inhabitants. Henry Slade, a clever American medium who was world-famous in the late nineteenth century, claimed that his controls had the power of moving objects in and out of 4-space during his séances. One of his favorite tricks was to produce knots in unknotted closed loops of rope, a feat that (barring trickery) could be explained only by assuming that part of the rope had been passed through a higher space. A German astronomer and physicist named Johann Carl Friedrich Zöllner, a remarkably stupid fellow who was incredibly ignorant of conjuring methods, fell completely for Slade's elementary brand of magic. Zöllner wrote an unintentionally hilarious book called *Transcendental Physics* in which he defended Slade's exploits against the charges of fraud.[4]

To obtain definitive, irrefutable proof of Slade's contact with spirits in 4-space, Zöllner once proposed that the medium reverse some dextrotartaric acid so that it would rotate a plane of polarized light to the left instead of the right. He also brought Slade a number of shells with conical helices that twisted right or left, to see if Slade could convert them to their mirror images. Such feats would surely have been as simple as tying a knot by passing part of a rope through 4-space, but from a conjuring standpoint they presented difficulties. Slade would have had to obtain some levotartaric acid, which could be synthesized only in a laboratory and was hard to come by, and it would have been even more difficult for him to find shells that were exact duplicates but of opposite handedness to the shells given him. As might be expected, neither of these crucial experiments succeeded. Of course, this made not the slightest dent in the hard shell of Zöllner's faith.

Is it possible that someday science will find evidence that a higher space is more than just a mathematical abstraction or the wild speculation of spiritualists and occultists? It is possible, though at present there are no more than tantalizing hints. The four-dimensional continuum of relativity is one in which 3-space is combined with time and handled mathematically as a non-Euclidian geometry of four dimensions. This is not at all the same thing as a 4-space consisting of four spatial coordinates. On the other hand, many cosmological models have been devised in which 3-space actually curves through 4-space in a way that could, in principle, be tested. Einstein, for instance, once proposed a cosmic model in which an astronaut could set out in any direction and if he traveled far enough, in the straightest possible line, he would return to his starting point. In this model our world of 3-space is treated as the hypersurface of an enormous hypersphere. Going around it would be comparable to a Flatlander's trip around the surface of a sphere.

In other cosmic models the hypersurface twists through 4-space in a manner analogous to such 2-space surfaces as the Klein bottle and the projective plane. These are closed, one-sided surfaces, without edges, which twist on themselves in a way similar to the way a Moebius strip twists. For example, if you suppose every point on a sphere is joined to every point exactly opposite it on the other side (you cannot imagine this; it has to be worked out mathematically), you have a model of what topologists call projective 3-space. An astronaut making a round trip through projective 3-space would return in reflected form, like H. G. Wells's Plattner.

To understand how the astronaut would be reversed, the following simple experiment is instructive. Cut two paper strips exactly alike, put one on top of the other, then (treating them as a single strip) make a half-twist and join the ends in the manner shown in Figure 44. The model you have formed is not the familiar Moebius strip, but the space *between* the two strips is.[5] The paper may be thought of as a covering for a Moebius surface of zero thickness. Now cut two small swastikas from a piece of dark-colored paper. Put both cutouts inside the double Moebius band, keeping them in place with paper clips as shown. The

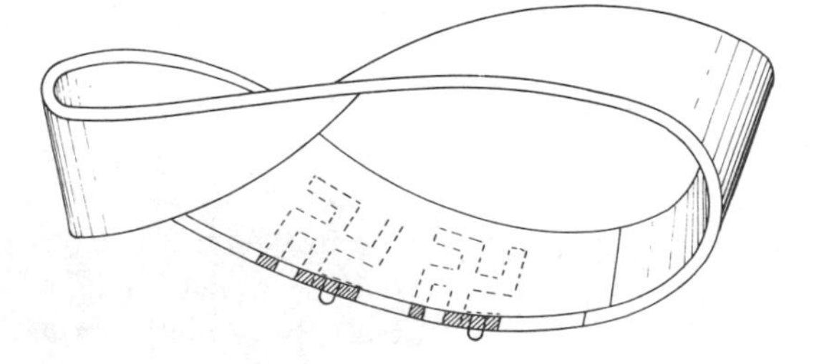

Figure 44. An experiment with a double Moebius band.

two swastikas must be placed side by side with the same handedness. Free one from its clip and slide it once around the Moebius surface, sliding it between the "two" strips until it is back where it was originally.

Examine the two swastikas. You will see at once that the cutout that made the round trip has changed its handedness. The two swastikas are no longer superposable. Of course, if you slide the cutout around once more it will recover its former handedness. This same sort of reversal would occur to an astronaut in 3-space if he made a round trip through a cosmos that twisted through 4-space in a manner analogous to the twist in a Moebius surface.

EXERCISE 12: *Figure 45 is a picture of a Klein bottle—a one-sided surface without edges. If an asymmetric Flatlander lived on such a surface (remember, it must be thought of as having zero thickness), would it be possible for him to make a trip around his cosmos in such a way that he would return in a form that was reversed with respect to his surroundings?*

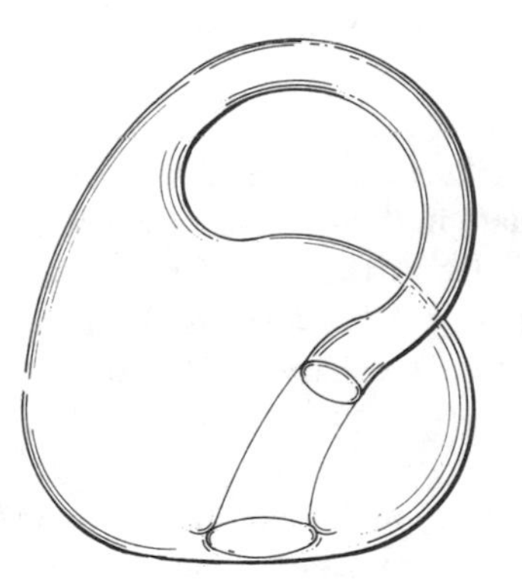

Figure 45. Model of a Klein bottle.

Notes

1. See Peter Remnant's paper on "Incongruent Counterparts and Absolute Space," *Mind*, vol. 73, July 1963, pp. 393–99, in which

Kant's thought experiment is analyzed, with conclusions essentially the same as those given here. For English versions of Kant's two early papers on space, see *Kant's Inaugural Dissertation and Early Writings on Space,* translated by John Handyside (Open Court, 1929). The thought experiment is discussed by Norman Kemp Smith in a section headed "The Paradox of Incongruous Counterparts," in *A Commentary on Kant's Critique of Pure Reason* (Macmillan and Co., Ltd., 1918, pp. 161–66); and in Hans Vaihinger's earlier German commentary on the same work, vol. 2, pp. 518*ff*.

2. These two lines are from Canto 2 of "Pale Fire," a beautiful poem by Vladimir Nabokov that is the heart of his remarkable novel of the same name. The poem is supposedly written by Nabokov's invented poet, John Francis Shade. As a joke, in the first edition of this book I credited the lines only to Shade and listed only Shade's name in the index. Nabokov returned the joke in his novel *Ada* (note the palindrome), where the action takes place on Anti-Terra, a kind of mirror image of our earth. On page 542 Nabokov repeats the same two lines, then adds that they were written by "a modern poet, as quoted by an invented philosopher ('Martin Gardiner') in *The Ambidextrous Universe. . . .*"

3. For two amusing recent stories about a man reversed in 4-space (both more up-to-date in their science than Wells's pioneer yarn), see "Technical Error" by Arthur C. Clarke (in Clarke's *Reach for Tomorrow,* Ballantine, 1956), and "The Heart on the Other Side" by George Gamow (in *The Expert Dreamers,* Frederik Pohl, ed., Doubleday, 1962).

4. Zöllner's book, first published in Germany in 1879, was later translated into English and issued in many editions. Sir Arthur Conan Doyle devotes a chapter to the defense of Slade in his *History of Spiritualism* (George H. Doran, 1926). A good discussion of Slade's methods of cheating will be found in section 2 of the *Proceedings of the American Society for Psychical Research, Inc.*, vol. 15, 1921, in an article by Walter F. Prince on "*A Survey of American Slate-Writing Mediumship.*" For more on this remarkable mountebank consult John

Mulholland, *Beware Familiar Spirits* (Scribners, 1938) and Harry Houdini, *A Magician Among the Spirits* (Harper, 1924).

5. Actually, there are not two strips but only one! For a discussion of some of the puzzling properties of this double Moebius band see chapter 7 of my *Scientific American Book of Mathematical Puzzles and Diversions* (Simon and Schuster, 1959).

18. THE OZMA PROBLEM

On controversial scientific questions for which there is a scarcity of empirical data, scientific opinion sometimes shifts back and forth like the changing fashions of women's clothes. The skirt is low in one decade, high in the next, then back down it goes again. When I was in college it was fashionable among astronomers to think that planets were extremely rare in the universe, on the theory that the earth was the result of an improbable collision or near approach of two suns. Quite possibly (it was believed) life in the cosmos is confined to our solar system, perhaps even to the earth. Today, informed opinion has swung the other way. Astronomers now suspect that planets are extremely *common* in the universe. Perhaps there are billions of them in our galaxy alone, millions of which may support intelligent life. If so, it seems likely that inhabitants of some of these planets, with a knowledge of science equal to or in advance of our own, may be trying to communicate with other planets.

On this assumption Project Ozma was started in 1960. A powerful radio telescope at Green Bank, West Virginia, was pointed toward various suns in the galaxy in a systematic search for radio messages from another world. Frank D. Drake, the radio astronomer who directed the project, is a long-time admirer of L. Frank Baum and his Oz books. He named the project for Ozma, the ruler of Baum's mythical utopia. It is an appropriate name. The location of Oz is unknown. Its inhabitants are "humanoid" but not necessarily "meat people" like us (witness the Tin Woodman and the Scarecrow.) Moreover, Oz is surrounded on all sides by the impassable Deadly Desert, which destroys anyone who so much as touches one grain of its sand. One of Baum's characters, the Nome King, has a servant called the Long Eared Hearer. The ears of this "nome" are several feet across. By placing one of them on the ground he can hear sounds thousands of miles away. Frank Drake's radio telescope is his Long Eared Hearer. It listened patiently for some type of coded signal, perhaps a repetition of a simple sequence of numbers, which could come only from an intelligent source that understood the universal laws of mathematics. The prospect of hearing such a signal is indeed an Ozzy one! It is hard to estimate the shattering effect such a signal would have on our self-centered, earthbound ways of thinking.

What should we do if we hear such a signal? Physicist Chen Ning Yang (we will hear more about him later) has made one suggestion: "Don't answer!" Such a response seems unlikely. Already, mathematicians and logicians are busy at work on step-by-step procedures by which two planets could slowly build up a common language for talking to each other. In 1962 Hans Freudenthal, a Dutch mathematician, published part 1 of an ambitious work called *Lincos: Design of a Language for Cosmic Intercourse.* There is no doubt whatever that coded pulses could be used for fluent communication. Once contact was made, it would be a simple matter to transmit detailed pictures. In crudest form it would only be necessary to divide a rectangle into thousands of tiny square units, like a sheet of graph paper, then transmit a binary code of ones and zeros indicating which unit squares—scanning the rectangle from top to bottom, left to right —should be blacked in. Better pictures, perhaps even moving TV pictures, could later be transmitted by the use of scanning beams. The long time intervals involved (it takes more than four years for a radio signal to reach the star nearest earth) introduce complications, but no one doubts that it would be only a matter of time until the two planets would be communicating with each other as easily, or almost as easily, as two nations on earth that speak different languages.

Did the reader notice the use of the phrase "left to right" in describing how that picture rectangle is to be scanned? Unless the inhabitants of the distant planet—we will call it Planet X for short—scan their rectangle from left to right, they will produce a picture which is a mirror image of the one we intend to transmit. How can we let them know what we mean by the phrase "left to right"?

Assume we have already established fluent communication with Planet X by means of a language such as Lincos and by the use of pictures. We have asked them to scan their rectangles from "top to bottom" and from "left to right." There is no possibility of their misinterpreting what we mean by "top to bottom." "Top" is the direction away from the center of a planet, "bottom" is toward the planet's center. "Front and back" is no problem either. But having established the meanings of up, down, front, back, how do we make clear our understanding of that third pair of directions, left and right? How can we be sure, when we transmit a picture of, say, what we call a right-handed helix, they receive a picture of a helix with the same handedness? If they have taken "left to right" in the same sense that we use the phrase, the pictures will match, but if they are scanning the other way, our picture of a right helix will be reproduced on Planet X as a left helix. In brief, how can we communicate to Planet X our meaning of left and right?

It is a puzzling question. Although an old problem, it has not yet been given a name.[1] I propose to call it the Ozma problem. To state it precisely: Is there any way to communicate the meaning of "left" by a language transmitted in the form of pulsating signals? By the terms of the problem we may say anything we please to our listeners, ask them to perform any experiment whatever, with one proviso: *There is to be no asymmetric object or structure that we and they can observe in common.*

Without this proviso there is no problem. For example, if we sent to Planet X a rocket missile carrying a picture of a man labeled "top," "bottom," "left," "right," the picture would immediately convey our meaning of "left." Or we might transmit a radio beam that had been given a helical twist by circular polarization. If the inhabitants of Planet X built antennas that could determine whether the polarization was clockwise or counterclockwise, a common understanding of "left" could easily be established. Or we might ask them to point a telescope toward a certain asymmetric configuration of stars and to use this stellar pattern for defining left and right. All of these methods, however, violate the proviso that there must be no common observation of a particular asymmetric object or structure.

Is it possible to transmit instructions for drawing a geometric design or graph of some sort that would explain to them what we mean by *left*? After considering it for a while, you can easily convince yourself that the answer is no. Every asymmetric pattern has both right and left forms. Until we and Planet X have a common understanding of left and right, there is no way to make clear which of the two patterns we have in mind. We could, for instance, ask them to draw a picture of a Nazi swastika, then define right as the direction toward which the top arm of the swastika points. Unfortunately, we have no way of telling them what we mean by a Nazi swastika. The swastika can spiral either way. Until we have agreed on left and right, we cannot give unambiguous instructions for drawing the swastika correctly.

Perhaps the field of chemistry would furnish a method of defining left and right. Could we explain to Planet X how to identify a crystal such as quartz or cinnabar that twisted polarized light a certain way? Yes, but even if they found such a crystal on their planet, the specimen would be of no help. As we learned in chapter 11, an optically active crystal can be of either handedness. Without a prior understanding of left and right, we would have no way of knowing the handedness of any particular crystal specimen they might find or grow in their laboratories.

The same ambiguity applies to all optically active stereoisomers. Every chemical compound capable of twisting polarized light—that is, every compound with atoms arranged asymmetrically in the molecule—also has both left and right forms. We could easily come to an understanding with Planet X about what we meant by an asymmetric form of tartaric acid, but if they succeeded in finding or synthesizing it we would not know whether they had obtained it in the right or the left form.

How about the asymmetry of carbon compounds in living tissues? We learned in an earlier chapter that all amino acids in living organisms on the earth are left-handed, and all helices of protein and nucleic acid are right-handed. If the inhabitants of Planet X are made of carbon compounds, perhaps they too contain protein and nucleic acid helices, and of course if they have proteins they also have amino acids.

Could we not define left and right in terms of the structure of such asymmetric carbon compounds? No, we could not. As we have seen, it is entirely accidental that our carbon compounds have their particular handedness. So far as we know there is no reason why every carbon compound in every living thing on earth could not, if evolution had taken a different turn at the beginning, have gone the other way. Without a prior understanding of right and left, we could not know whether *their* amino acids were right- or left-handed.

Assume that their planet, like earth, is rotating on an axis. Is there any way this rotation could be used as a basis for defining left? The direction of rotation of the earth can be demonstrated by means of a heavy weight suspended by a long fine wire and swinging slowly back and forth. The device is known as a Foucault pendulum, after Jean Bernard Léon Foucault, the French physicist who first demonstrated it, in Paris in 1851. The swinging weight's inertia keeps the direction of its swing constant in relation to the stars while the planet rotates beneath it. In the Northern Hemisphere a Foucault pendulum rotates clockwise; in the Southern Hemisphere it rotates the other way. But how could we explain to Planet X what we mean by North and South Hemispheres? We could not say: Stand on your equator, facing the direction your planet rotates, and the Northern Hemisphere will be on your left. That would presuppose an understanding of "left." Unless we could make clear to Planet X which hemisphere was which, the Foucault pendulum would be no help. The same is true of the various asymmetric phenomena that are the result of a planet's Coriolis forces. We could not say: Fire a missile from the equator toward your North Pole and you will see it deviate in the direction we call "right." Such a statement would be ambiguous unless we had previously agreed on which pole was "north." This we could not do without an agreement on what we meant by left and right.

Perhaps Planet X has a magnetic field with north and south poles that correspond closely to the poles of the planet's axis of rotation. Would that be of any help? No. In the first place, we do not know yet the cause of a planet's magnetic field. Presumably it is related in some way to a planet's rotation, but we cannot say with assurance that what we call a north magnetic pole is always associated with the end of the axis of rotation that is on the left when you face the direction of rotation. It may be on the right. The sun always rotates in the same direction, but as we learned in chapter 6, every now and then the magnetic poles of the sun do a peculiar flip-flop; the north pole becomes the south pole and vice versa. The moon, which rotates slowly (one rotation for each revolution around the earth), apparently has *no* magnetic poles. We have no grounds, at present, for guessing how the magnetic poles of Planet X would be placed with respect to the direction of the planet's rotation. Even if we *did* know how they were placed, it still would not help us define left and right, as we will see in the next chapter.

One possibility remains: the asymmetric phenomena associated with electrical and magnetic forces. To take the most familiar example, the magnetic lines of force surrounding a current go around the current in a counterclockwise direction if you face the direction of current flow. In the nineteenth century, when it was thought that current flowed through a wire from positive to negative poles of a battery, this asymmetry was expressed by what physicists called the right-hand rule. If you grasped a wire with your right hand, its thumb pointing along the wire from positive to negative poles, your fingers would curl around the wire in the direction of the magnetic lines of force. Today we know that the current actually flows the opposite way. The motion of free electrons, which produces the wave pulse that is the electric current, goes from the negative pole of a battery to the positive. In this book we adopt the practice of physicists who prefer the convention of a "left-hand rule."

Exactly what does a physicist mean when he says that if you curl your left fingers around a wire, thumb pointing in the direction of current flow, the fingers will point in the direction of the current's magnetic field? He means that if you put a magnetic needle near the wire, the north pole of the needle will always point in a direction counterclockwise around the wire as you face the direction of current flow. Figure 46 shows how the magnetic needle behaves when placed at various positions around a wire carrying a current moving in the direction of the arrow.

Here we have a simple, striking instance of asymmetry. We could explain to the inhabitants of Planet X exactly how to make a battery by mixing certain chemicals and inserting metals in the liquid to provide positive and negative poles. Once we and planet X agreed on the direction of current flow along a wire (there is no difficulty in

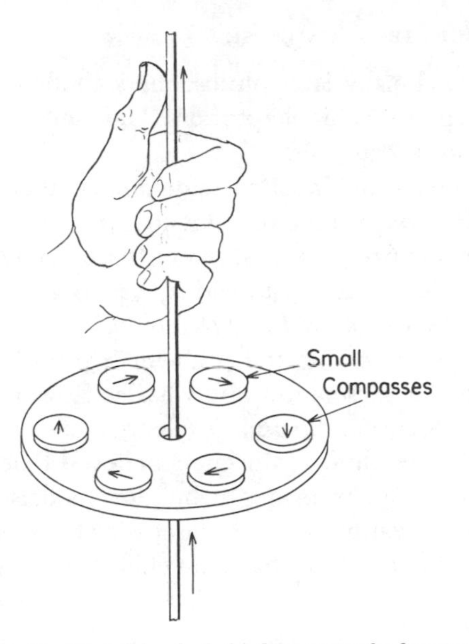

Figure 46. The left-hand rule for determining the direction of a magnetic field surrounding an electrical current.

agreeing on this) could we not then say: Put a magnetic needle above the wire, face the direction the current moves, and the north end of the needle will point in the direction that we on earth call left?

Here, surely, is a simple experiment that provides a clear, unambiguous, operational definition of left and right. No?

No. The experiment would do the trick only if we had some unambiguous way of telling Planet X which end of the needle is the end we call north. Alas, there is no way of communicating this necessary information without first having a common understanding of left and right. To understand why this is so, we must first understand the fundamentals of the modern theory of magnetism. This will be the task of the following chapter.

Notes

1. I do not know who was the first to give this problem explicitly as one of communication. It is, of course, implied in Kant's discussion of

left and right, and many later philosophers allude to it. This is how William James puts it in his chapter on "The Perception of Space" in *Principles of Psychology*, 1890:

"If we take a cube and label one side *top*, another *bottom*, a third *front*, and a fourth *back*, there remains no form of words by which we can describe to another person which of the remaining sides is *right* and which *left*. We can only point and say *here* is right and *there* is left, just as we should say *this* is red and *that* blue."

James's way of presenting the problem is probably based on his reading of a similar presentation by Charles Howard Hinton in the first series of his *Scientific Romances* (George Allen & Unwin, 1888). Hinton (we will meet him again later) believed that he had taught himself to think in 4-space images by building models with cubes that had been colored in various ways. In discussing these cubes (page 220) he gives a clear statement of what I am calling the Ozma problem.

19. MACH'S SHOCK

Picture a wire running north and south beneath a compass (Figure 47, left). The compass needle parallels the wire and points north. Now an electrical current is sent through the wire from south to north. The needle immediately swings counterclockwise and points due west

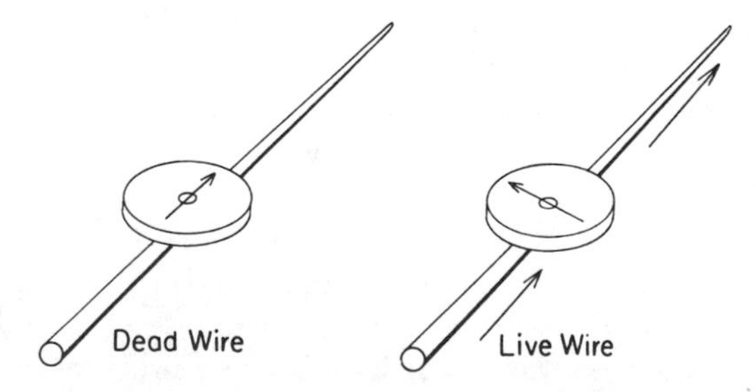

Figure 47. The experiment that "shocked" Mach.

(Figure 47, right). If the direction of current is reversed, the needle does an about-face and points due east.

Nineteenth-century physicists assumed that this indicated some sort of mysterious asymmetry in the laws of nature. The experiment was not superposable on its mirror image, for in a reflection of the experiment the north end of the compass needle would point the wrong way. The great German physicist Ernst Mach, in his classic *The Science of Mechanics,* emphasized the "intellectual shock" produced by this simple experiment.[1] It teaches us, he said, an important lesson. We must always be suspicious of our intuitions when we try to guess in advance how nature will behave.

A man said to the universe:
"Sir, I exist!"
"However," replied the universe,
"The fact has not created in me
A sense of obligation."[2]

The universe is under no obligation to conform to any scientist's desires or intuitive guesses. In the needle experiment our intuitions lead us to expect that electrical and magnetic fields, like the symmetric fields of other physical forces, will show no bias for right or left. Yet some sort of asymmetric twist seems to be an essential part of the wire and compass experiment.

Will not this twist provide a simple basis for defining left and right, and consequently solve the Ozma problem? We have only to ask our friends on Planet X to set up the experiment, then we all agree that "left" is the direction the compass needle points when the current beneath it is moving away from us. Where is the flaw in this procedure?

The flaw lies in our curious inability to communicate to Planet X which pole of the needle is to be labeled "north." If all north poles of magnets were red and all south poles green, there would be no difficulty. We could tell Planet X that the north pole was the red pole. Unfortunately, no amount of inspection or testing of a magnet discloses the slightest difference between the poles. Their strengths are precisely the same. A magnetized needle floating on water shows no tendency to drift either north or south. If the surface of a bar magnet is highly polished and coated with a liquid containing an iron powder, the particles of iron form "domain" patterns on the magnet (we will explain domains in a moment), which can be seen with a microscope. But the patterns show no bias toward either end of the magnet; they provide no clue for distinguishing between the two poles. Every now and then, during the past fifty years, a physicist has thought he has discovered some intrinsic feature by which one magnetic pole could be distinguished from the other without testing the poles on outside magnetic fields. Sometimes papers reporting such "discoveries" are published in physics journals. It always turns out that the physicist is mistaken.

The north end of a compass needle is usually painted black to distinguish it from its south end. How does a compass maker know which end to paint black? By testing it with other magnets. The north end is the end that is repelled by the north poles of other magnets. And how does one recognize the north poles of other magnets? They are the ends that are repelled by the north poles of still other magnets. The ultimate basis for the definition of "north pole" is the magnetic field of the earth itself. A magnet's north pole is the pole attracted by the earth's north magnetic pole.

This is somewhat confusing because like poles repel each other. Strictly speaking, the earth's north magnetic pole is its "south" pole. But convention has it that the earth's south magnetic pole, because of its nearness to the earth's geographic North Pole, be called the north magnetic pole. The important point is that we have no way of telling Planet X which end of a magnetized needle is the end we call north because we have no way of telling them which end of the earth's axis of rotation is the end we call north.

If a wire is wrapped around an iron or steel core and a current is passed through the wire, the core becomes an electromagnet. It is possible to wrap the wire in such a way that the north pole of the electromagnet can be placed at either end of the core. Could we not send Planet X instructions about how to make an electromagnet that could then be used for establishing an unambiguous definition of a magnet's north pole?

Readers acquainted with elementary physics will shake their heads at once. The wire that coils around the core of an electromagnet forms either a right- or left-handed helix. If the current travels around the core in a right-handed helix, it will travel toward the core's south pole

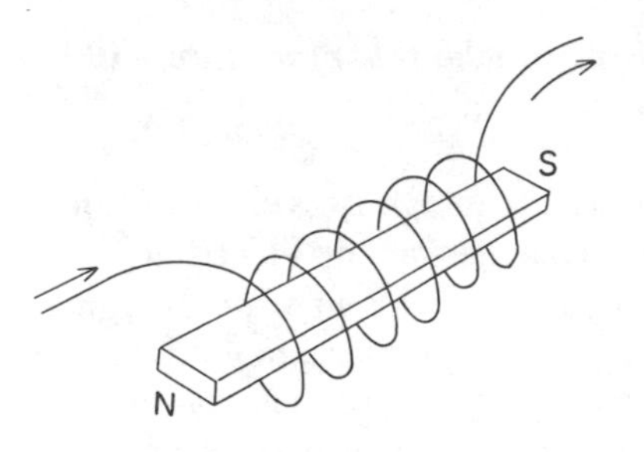

Figure 48. An electromagnet.

(see Figure 48). If it moves in a left-handed helix, it travels toward the core's north pole. Even without a core, the helical wire has a magnetic field with north and south ends. In both cases the left-hand rule determines which end is which. If you put your left hand on the helix, your fingers curving in the direction the current flows around the coil, your thumb will point toward the north pole of the field produced by the helix. Clearly, we cannot explain to Planet X which end of an electromagnet is north until we can explain what we mean by a right-handed helix. This we cannot do without a prior understanding of left and right.

George O. Smith's "Amateur in Chancery" (*Galaxy*, October 1961; reprinted in Frederik Pohl's anthology, *The Expert Dreamers*, Doubleday, 1962) is a science-fiction story based on the difficulty of communicating to a Venusian our meaning of right and left. Someone in the story suggests the following procedure:

"Let's wind our electromagnet like this: We place the steel bar horizontally in front of us. The wire from 'Start' leaves us, passes over the top of the bar, drops below the bar on the far side, comes toward us on the under side, rises above the bar on the side toward us, and so on around and around until we've got our electromagnet wound."

There are two ways of following these instructions. If the wire is wound with the right hand, it forms a left-handed helix around the steel bar. If wound with the left hand, it forms a right-handed helix. In *both cases*, however, if current enters the wire from the "Start" end, application of the left-hand rule shows that the electromagnet's north pole will be on the right. The direction of current flow *can* be communicated.

EXERCISE 13: *Explain why this procedure will not communicate the meaning of left and right.*

A similar left–right ambiguity is involved in all asymmetric phenomena related to magnetism and electricity. Just as moving electric charges (currents) create fields in which magnets orient themselves asymmetrically, so do magnets create fields in which currents tend to behave in a seemingly asymmetric fashion. A well-known experiment is one in which a vertical wire, its tip immersed in mercury, is made to circle around the pole of a magnet in either a clockwise or counterclockwise direction. The same principle underlies a primitive type of motor known as Barlow's wheel. In all such phenomena the direction of rotation depends on which pole of the magnet is used. We cannot use these rotations for communicating left and right to Planet X because we cannot tell them how to distinguish the north from the south pole of a magnet.

Similar ambiguities surround the asymmetric motions of charged particles in magnetic fields. A charged particle that acquires a right-handed helical path in moving through a magnetic field will acquire a left-handed helical path if the poles of the field are reversed. No experiment with electric charges and magnetic fields can provide an unambiguous definition of left and right. At some point in the experiment there is introduced either a left–right distinction or a distinction between north and south magnetic poles that in turn rests on a left–right distinction.

Physicists like to put it this way: The difference between the north and south poles of a magnetic field is a matter of convention. We know that unlike poles attract, like poles repel, so it is necessary that the poles have different names. We call one pole north because it is the pole attracted by the earth's north (really its south) magnetic pole. We call the other pole south because it is the pole attracted by the earth's south (really its north) magnetic pole. These are no more than names introduced for the sake of convenience. The magnetic field of a bar magnet is absolutely symmetrical with respect to a plane that cuts the polar axis in the middle. If suddenly the north pole of every magnetic field in the universe became a south pole, and every south pole became a north, there would be no change that could be detected by any experiment. It would be as meaningless to say such a change had occurred as it would be to say the cosmos had turned upside down. (This is what twentieth-century physicists would have said prior to 1957. In 1957 something happened that changed the picture radically, but we don't want to get ahead of our story.)

The situation continues, however, to be puzzling. After all, a magnetic needle *does* behave in a strangely asymmetric way when placed above or below a current. True, we cannot examine the ends of a magnetized needle under a microscope and find anything to tell us which pole is which. Nevertheless, one pole clearly *is* north and the other south. There clearly is *some* difference between the poles, otherwise why would unlike poles attract and like poles repel? If we paint the north pole of a magnetized needle red, it will always be the red end that points left when we place it above a current moving away from us. How can one explain this seeming asymmetry—this "Mach shock"—and still maintain that electromagnetic fields are fundamentally symmetrical?

The full answer did not come until the twentieth century when physicists discovered that the properties of a magnet were simply the consequences of rotary movements of charged particles within the magnet. To make this clear we must first take a quick look at the structure of atoms. The look will be at what is called the Bohr model of the atom, a model based on the theoretical work of Niels Bohr, the great Danish physicist who died in 1962. This model is now known to be only a crude approximation. It is (as George Gamow has put it with his usual aptness of imagery) the atom stripped of its flesh until only its skeleton remains. The full, fleshy details can be described accurately only by the complicated mathematics of modern quantum theory. The Bohr model, nevertheless, is still enormously useful in giving a rough, symbolic picture of what is known about the atom's structure; there is no reason why we should be ashamed to use it.

In the Bohr model of the atom, the nucleus has one or more electrons traveling around it in orbits arranged in shells. Each electron carries a single charge (a quantum) of negative electricity. Normally the atom is in an uncharged state with the number of electrons balancing the number of protons in the nucleus. Each proton carries a quantum of positive charge. In addition there may be one or more neutrons (uncharged particles) in the nucleus.

Figure 49 is a picture of the simplest of all atoms, an atom of

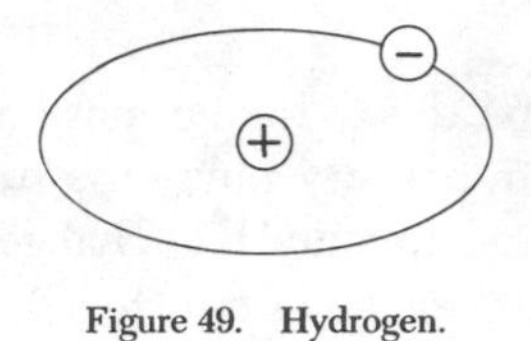

Figure 49. Hydrogen.

hydrogen. The nucleus consists of one positively charged proton. Around it circles one negatively charged electron. If the nucleus has, in addition to the proton, a single neutron, then it is one of the *isotopes* of hydrogen (see Figure 50). (An isotope is a variant form of an

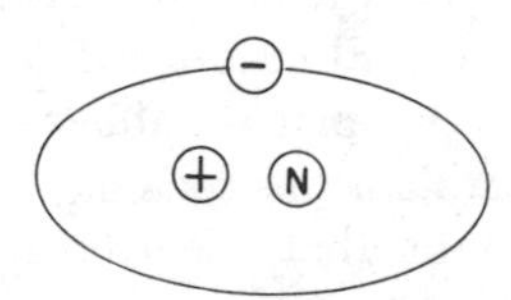

Figure 50. Deuterium, or heavy hydrogen.

element that results when the nucleus varies in its number of neutrons.) This isotope is called deuterium, because it has two particles in its nucleus. The added neutron makes it heavier, and for this reason it is often called heavy hydrogen.

Figure 51 is a picture of the next simplest atom: helium. In its most common form it has a nucleus containing two protons and two neutrons. Whirling around the nucleus are two electrons.

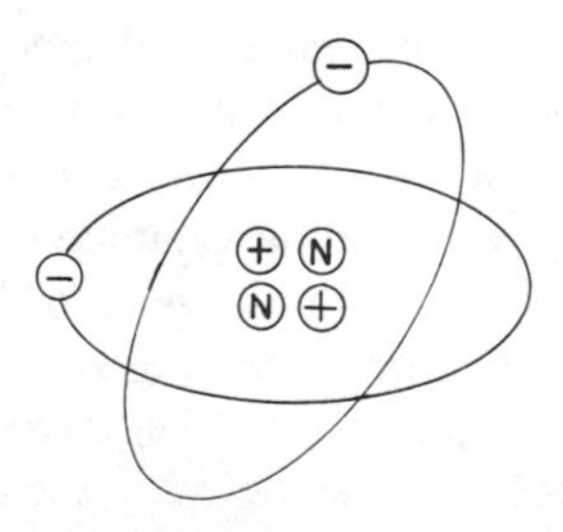

Figure 51. Helium.

Because the structure of an atom is roughly spherical, it is convenient to picture the atom in the mind as a tiny ball. "To some teachers an atom is always a ball," declared nuclear physicist Samuel Goudsmit (as quoted in Daniel Lang's book *The Man in the Thick Lead Suit,* 1954). "In the winter it's a basketball, in the spring it's a baseball, and the rest of the time it's a ping-pong ball. The atom is no more explained by such images than the idea of God is by a picture of an old man with a long beard sitting on a cloud."

It is good to be reminded of the crudity of models. On the other hand, it would be difficult to get along without them. Chemists still diagram molecules by using dashes to symbolize complicated valence bonds holding the atoms together, and for similar reasons physicists continue to talk in terms of the Bohr model. It is a handy symbolic shorthand. Why *shouldn't* the atom be called a ball? After all, what is a ball? In ordinary language it is any object of roughly spherical shape. A term loose enough to include a football, popcorn ball, and a handkerchief wadded into a ball is surely loose enough to cover the fuzzy ball-like structure of an atom, even though its "cloud" of electrons can be described only by complicated probability functions.

The electron that circles the nucleus of an atom is a moving charge of negative electricity. Its motion creates a magnetic field with an axis through the atom's center, perpendicular to the plane of the electron's orbit. This field is called the *orbital magnetic moment* of the electron. In addition to its orbital motion, the electron also has what is called *spin.* (Dr. Goudsmit, quoted above, is one of the codiscoverers of spin.) In the Bohr model, spin may be pictured as the rotation of the electron about an axis through its center, just as the earth rotates as it goes around the sun. This also creates a tiny magnetic field with an axis coinciding with the axis of rotation. It is called the spin magnetic moment of the electron. The existence of this field suggests that an electron must have a structure that actually moves through space, but if so, what that structure is remains a profound mystery.

Figure 52 shows the magnetic axis of an electron's orbital magnetic field. The end labeled *north* is the end from which, if you look down on the electron, you see its path as clockwise around the nucleus. Figure 53 shows the magnetic axis of an electron's spin magnetic field. Again, the north end is the end from which, if you stood on it and looked down, you would see the electron spinning clockwise. In both cases the labeling is such as to conform to the conventional use of the left-hand rule. Physicists prefer to use a plus sign for north and a

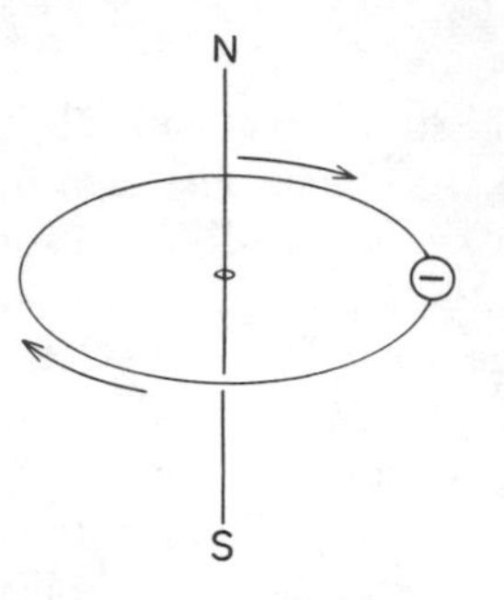

Figure 52. Orbital magnetic moment of electron.

minus sign for south, but since this is not a technical book we will stick to the more familiar terms.

In addition to magnetic fields generated by the spins and orbital motions of electrons, there are similar fields generated by spins of protons, neutrons, and even by the spin of the atom's nucleus as a whole. (Why the spinning neutron, which has no electric charge, would create a magnetic field is still something of a mystery. We will consider it again later.) That the term *spin* is an appropriate one is shown by the fact that particles with spin actually behave like tiny gyroscopes; they resist turning motions. Many laboratories are at work developing nuclear gyroscopes for guiding spaceships; these fantastic gyroscopes have no moving parts to wear out and show no drift as a result of friction. They are based on the gyroscopic properties of spinning nuclear particles.

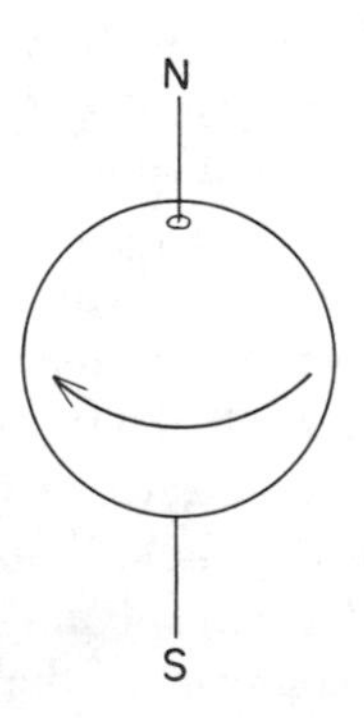

Figure 53. Spin magnetic moment of electron.

When any two magnetic moments within an atom have their axes parallel or nearly parallel, with their north ends pointing the same way, the fields combine to produce a stronger field. If the axes are antiparallel (point in opposite directions) the fields neutralize each other to produce a weaker field or no field at all. For example, the two electrons in the helium atom revolve in the same orbit but in opposite directions; therefore their orbital magnetic moments cancel out. The same is true of their spin magnetic moments. One spins clockwise, the other counterclockwise. The atom is said to be *spin balanced.* This mutual annihilation of orbital and spin magnetic fields results in a helium atom that is magnetically neutral. It has no overall or *resultant* magnetic moment. The same is true of all other rare gases (neon, argon, krypton, xenon, radon)—gases that have outer shells filled with their full quota of electrons. Other atoms, because of a lack of balance of their internal magnetic moments, have an overall magnetic field. (In more technical language, the resultant field is the vector sum of all internal magnetic moments.) Such an atom can be said, in symbolic terms, to possess an overall spin that gives it a resultant magnetic field with north and south poles. In brief, it behaves like a tiny spherical magnet.

Among the atoms of all the elements, the iron atom (because of a great imbalance of electron spins) has the strongest resultant magnetic field. We can think of each atom in a bar of iron as a tiny spherical magnet with north and south poles. Every atom is firmly locked into position in the cubical lattice of the iron crystal but free to turn so that its magnetic axis can point in different directions. Magnetizing an iron bar is nothing more than causing its atoms to turn until as many as possible are lined up with the magnetic axes parallel. Because parallel magnetic moments reinforce each other, the bar acquires a strong resultant field.

There is a limit, of course, to the field's strength. The atoms of an unmagnetized iron bar are like a crowd of people packed into a room and facing different directions. The room is "magnetized" by a magnetic orator who persuades as many people as possible to face toward him. The more turn to face him, the stronger the resultant magnetic field. The field reaches its saturation point when everyone in the room faces the same way. Obviously, there is no way to make the field any stronger.

For complicated reasons that cannot be gone into here, the atoms in an unmagnetized bar of iron are not oriented in individually random ways. They tend to form little clumps or sets called domains in which the magnetic axes of the atoms are parallel. It is these domains, not the individual atoms, that have their magnetic axes turned in different directions. When the bar is magnetized by placing it in a strong outside field, the walls of these little domains shift as their atoms turn to align their axes with the axis of the outside field.

Many elementary physics textbooks, especially those published before 1950, give a false impression of what happens when a bar of iron is magnetized. One picture will show the domains inside an unmagnetized bar as tiny little bar magnets pointed in random directions. Beside it is shown a magnetized bar with these little magnets lined up and all pointing the same way. One gets the impression that the domains are rigid little pieces of iron inside the bar that actually swing around when the bar is magnetized. This could not be the case, because each iron atom is locked permanently in place in the iron's lattice.

Think of a large company of soldiers standing in a square lattice formation on a large field. Each soldier is firmly rooted to a spot on the ground but free to turn in different directions. A formation of eighteen soldiers, standing three in a row, is facing north. Behind them a group of eighteen soldiers, also three in a row, is facing south. Each group represents a domain of iron atoms. Imagine now that the domain of south-facing soldiers is persuaded to face north. Instead of turning simultaneously, however, first the northernmost row of soldiers turns, then the next row, then the next, until finally all the soldiers in the domain are facing north. As the rows turn, the domain "wall"—the dividing line between the two sets of soldiers—moves gradually south until the two domains coalesce into one large north-facing domain. This gives a rough picture of what happens to the atoms in an iron bar as the bar is being magnetized.

The domains in the bar do not all swing into line simultaneously. As a result, the bar's magnetic field grows stronger in a series of abrupt little jumps. If a wire is wrapped around a bar that is being magnetized, each jump induces a small voltage in the wire. These electrical impulses can be amplified and actually heard as a series of clicks—a sort of rustling sound like that of paper being crumpled. This is called the Barkhausen effect, after Heinrich Barkhausen, a German engineer who discovered it in 1919. If you should visit the fabulous Museum of Science and Industry in Chicago, you can push a button and actually hear the Barkhausen effect. As you watch a small bar of iron move slowly into a magnetic field, you will hear the amplified rustling noise that results from the discontinuous movement of domain walls as the bar's atoms swing into alignment.

For many centuries physicists were puzzled by the fact that it is impossible to create a *monopoled* magnet: a magnet with only one pole. They were puzzled also by the fact that, whenever a bar magnet is cut in half, it always results in two smaller bar magnets. If the two halves are cut in half, there are four little magnets, each complete with a north pole at one end and a south pole at the other.

The modern theory of magnetism completely clears up both mysteries. Think of a bar magnet symbolically as a cylinder with little arrows painted on it as shown in Figure 54. The arrows indicate the direction in which the majority of the bar's electrons are spinning. It is this overall spin of the cylinder that makes it a magnet. If you look at one end of the cylinder you will see this spin as clockwise. By convention, this end is called the cylinder's north magnetic pole. Look at the other end and you will see the spin as counterclockwise. That end is the south magnetic pole. The poles are no more than labels for the enantiomorphic opposite ends of a (symbolically) rotating cylinder. They are certainly not "things" in the sense that positive and negative charges are "things." (Perhaps we should say they are not entities in the same sense that positive and negative charges *seem* to be entities in the light of our present ignorance of what they are.)

It is easy to see why there are no monopoles[3] and why any segment cut from a bar magnet cannot fail to have north and south ends. It

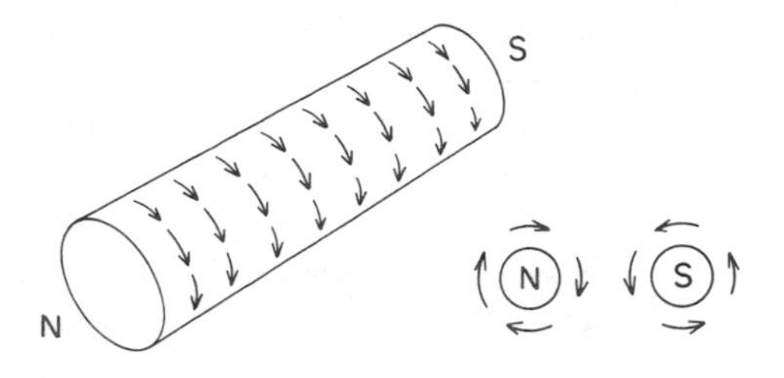

Figure 54. Symbolic picture of a bar magnet.

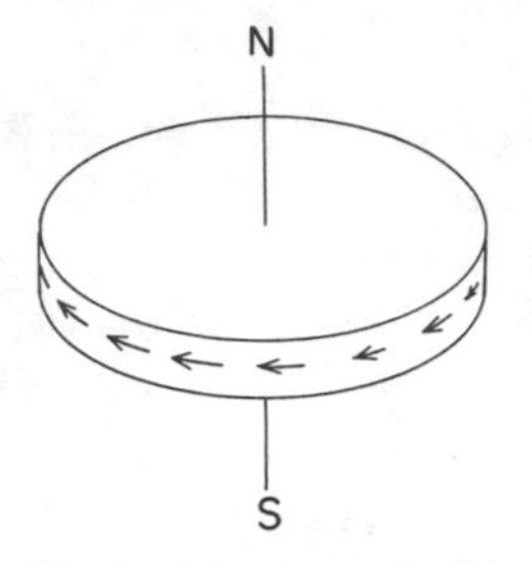

Figure 55. Symbolic picture of a disk magnet.

would be as difficult to have a monopoled magnet as it would be to spin a cylinder that had only one end. Even a disk-shaped magnet such as shown in Figure 55, with its magnetic axis perpendicular to the surface, must be north on one side, south on the other. It would be as difficult to make a disk magnet with north on both sides as it would be to spin a wheel in such a way that you could see it as spinning clockwise from both sides. It would be as impossible to cut a bar magnet in half and not produce two smaller replicas as it would be to cut a spinning cylinder in half and not produce two smaller spinning cylinders.

We are finally in a position to understand why magnetic field reactions do not represent any basic departure from symmetry. Think of all the magnetic fields in the universe as cylinders of various sizes, from the size of an electron to the size of a galaxy, each painted with arrows to indicate the direction of spin. Hold such a cylinder up to a mirror; you see at once that it can be superposed on its reflection. All you have to do is turn the mirror image around and its arrows will coincide with those on the actual cylinder. If one end of the cylinder differed in any essential way from the other—if, for example, the cylinders were cones—they would then be asymmetric and not superposable on their reflections. But the ends do not differ.

The fact that spinning cylinders are symmetrical does not, of course, prevent their ends from coming together in two essentially different ways. If they come together so that their spins (arrow directions) go the *same* way, there is a meeting of *opposite* poles. The spins reinforce each other, and there is a strong attraction between them. If they come together with their spins (arrow directions) *opposing* each other, there is a meeting of *like* poles. The spins counter each other, and there is a strong repulsion. For convenience it is necessary to give different labels to the two ends. Once we have decided on a label for one end, we automatically determine the label for all the ends on all the cylinders (magnetic fields) in the universe. As we have seen, scientists decided to call north that end of a bar magnet that was attracted by what we call the north magnetic pole of the earth. Once this decision was made, every magnetic field in the cosmos acquired a labeling of ends that would conform to this initial choice.

Do you see now why we cannot use the needle and wire experiment (or any similar instance of magnetic asymmetry) for communicating our meaning of left and right to Planet X? We can tell them to suspend a bar magnet above a current. We can explain to them how the bar behaves like a cylinder with arrows painted on it, turning until the arrows nearest the wire are pointing in the same direction the current is moving (Figure 56). Now we are hopelessly stuck. Since the cylinder's two ends are alike in every respect except that one is a mirror reflection of the other, we have no way of telling Planet X which end we have decided to call north. We can say, "North is the end of the cylinder with arrows that go clockwise when you look at the end," but we have no way of explaining what we mean by "clockwise." Magnetism is no more help in solving the Ozma problem than the existence in the world of rotating wheels and cylinders. What seemed to Mach, Pasteur, and other scientists of the time to be a clear case of asymmetry in natural law proves to be pseudo-asymmetry once the modern theory of magnetism is understood.

The experiment that shocked Mach can be likened to a row of hamsters, joined side by side, trotting on a wide treadmill (see Figure

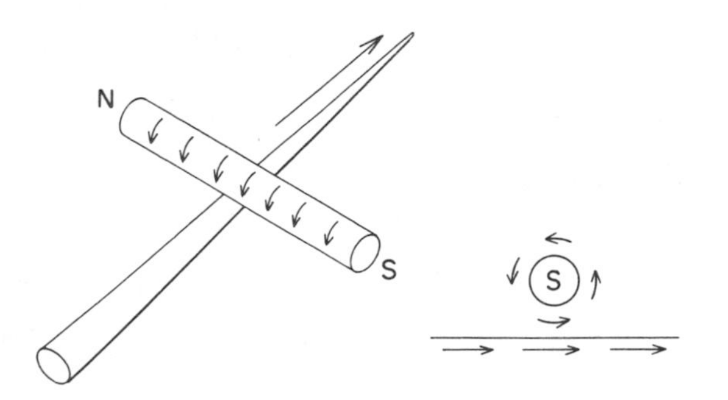

Figure 56. Symmetry in the wire and needle experiment.

57). A motor turns the treadmill so that its upper surface moves from south to north. This motion corresponds to the northward motion of the wire's current. The hamsters are the atoms in a magnetized bar of iron. Because the little animals find it difficult to trot backward or sideways, they naturally turn so that they all face south. The end of the row on their right—the end that points west—corresponds to the north pole of the bar magnet. The end of the row that points east is the south pole.

What happens if you lift up the entire row of hamsters and replace them on the moving treadmill so that their north pole points north, with all the animals facing west? The entire row will execute a left-face in order to return to its original position. The hamsters will never execute a right-face, because if they did they would find themselves forced to trot backward on the treadmill. Could this situation be used for communicating to Planet X an operational definition of left and right? No, because the row of hamsters is bilaterally symmetric. To make clear to Planet X which end of the row always points left, as the observer faces the direction of the treadmill's motion,we would first have to make clear what we mean by the right and

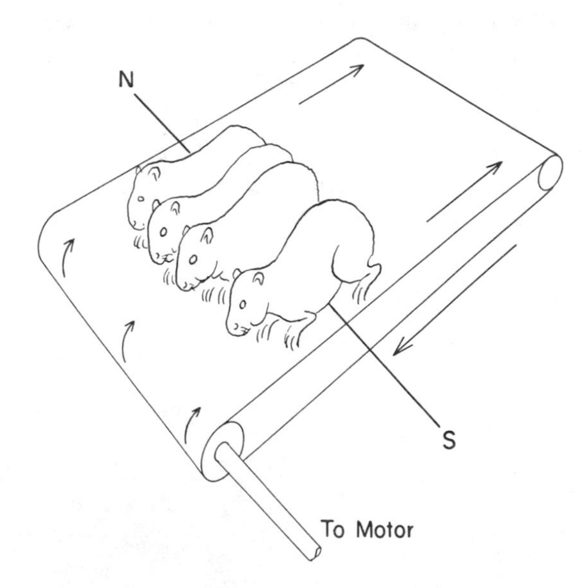

Figure 57. Symbolic picture of the wire and needle experiment.

left sides of a hamster. This of course is precisely what we have not yet found a way to do.

Mach's intuitions were better than he knew. The magnetic field turns out to be symmetrical after all! It was not until 1957 that an experiment much more shocking was announced, but we have many preliminaries yet to clear away before this experiment can be described.

Notes

1. "Even instinctive knowledge of so great a logical force as the principle of symmetry employed by Archimedes, may lead us astray. Many of my readers will recall to mind, perhaps, the intellectual shock they experienced when they heard for the first time that a magnetic needle lying in the magnetic meridian is deflected in a definite direction away from the meridian by a wire conducting a current being carried along in a parallel direction above it. The instinctive is just as fallible as the distinctly conscious. Its only value is in provinces with which we are very familiar." Ernst Mach, *The Science of Mechanics,* translated by Thomas J. McCormack (Open Court, 1893), chapter 1.

2. From Stephen Crane, *War Is Kind and Other Lines* (Knopf, 1899).

3. The reference here is to monopoled magnets, not "magnetic monopoles," which may or may not exist. In 1931 P. A. M. Dirac conjectured that there is an elementary particle carrying a quantum of either north or south magnetic charge. He called them monopoles. If they exist they would introduce a beautiful symmetry into the equations for electromagnetism. Just as a moving unit of electric charge (such as the electron has) creates a magnetic field, so a moving unit of magnetic charge would create an electrical field. Like electric charges, monopoles could be created only in pairs of opposite sign, and if two opposite monopoles met they would annihilate each other.

In Dirac's theory a monopole's charge has to be an integral multiple of 68.5, or half of 137, the reciprocal of the mysterious fine-structure constant. Physicists find it hard to explain why such monopoles do not exist, but so far no monopole has yet been observed. There was a flurry of excitement about catching one in 1975, but the consensus now is that what was caught was something else. See Kenneth E. Ford, "Magnetic Monopoles," *Scientific American,* December 1963.

20. PARITY

Twenty years ago, had you asked a physicist for a solution to the Ozma problem, you would have been told: There *is* no solution. There is no way, he would have said, to communicate the meaning of left and right to the intelligent beings on Planet X without turning their attention toward a particular asymmetric structure—a configuration of stars, a beam of circularly polarized light, or the like—which both we and they could observe in common. There is no experiment, involving any of the known laws of nature, that can provide an operational definition of left and right.

When something in nature always remains the same, physicists like to express the invariance by a conservation law. For example: the law of the conservation of mass-energy states that the total amount of mass-energy in the universe never changes. Mass may change to energy and vice versa (in accordance with Einstein's famous formula, $E = mc^2$), but there is never an increase or loss in mass-energy. The conservation law that implies the universe's fundamental, never-changing mirror symmetry—its lack of bias for left or right in its basic laws—is the law of the conservation of parity.

The term *parity* was first used by mathematicians to distinguish between odd and even numbers. If two integers are both even or both odd, they are said to have the same parity. If one is even and the other odd, they are said to have opposite parity. The term came to be applied in many different ways to any situation in which things fall neatly into two mutually exclusive classes that can be identified with odd and even integers. For a simple illustration, place three pennies in a row on the table, each head-side up. Now turn the coins over, one at a time, taking them in any order you please, but make an *even* number of turns. You will find that no matter how many turns you make—2, 74, 3,496, any even number—you are sure to end with one of the following four patterns:

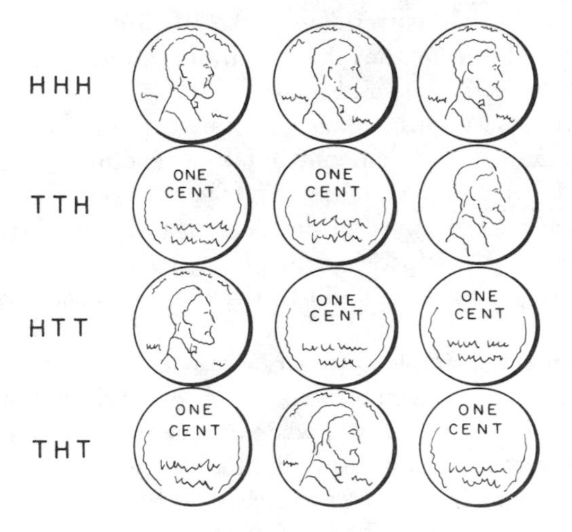

Place the three pennies, all heads up, in a row again. This time make an *odd* number of turns, taking the coins in any order you please. You are sure to end with one of the next patterns shown.

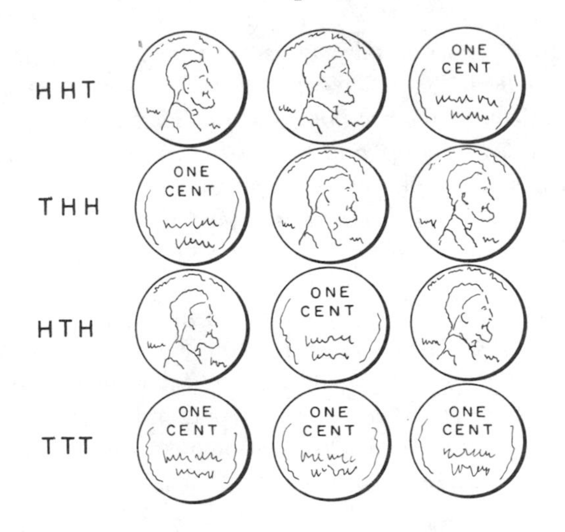

The first set of patterns can be said to have even parity, the second set an odd parity. Experiment will show that the parity of a pattern is conserved by any even number of turns. If you start with an even pattern and make, say, 10 turns, the final pattern is sure to be even. If you start with an odd pattern and make 10 turns you are sure to end with an odd pattern. On the other hand, any pattern changes its parity if you make an odd number of turns.

Many tricks with cards, coins, and other objects exploit these principles. For example, ask someone to take a handful of coins out of his pocket and toss them on a table. While your back is turned, he turns over coins at random, one at a time, calling out "Turn" each time he reverses a coin. He stops when he pleases, covers one coin with his hand. You turn around and tell him whether the hidden coin is heads or tails.

The method is a simple application of what mathematicians call a "parity check." Before you turn your back, count the number of heads and remember whether it is an even or odd number. If he makes an even number of turns you know that the parity of the heads remains the same. An odd number of turns changes the parity. Knowing the parity, a simple count of the heads showing, after you turn around, will tell you whether the hidden coin is heads or tails. To vary the trick, you can have him cover two coins and tell him whether they match or not.

EXERCISE 14: *Place six drinking glasses in a row, the first three brim up, the next three brim down. Seize any pair of glasses, one in each hand, and simultaneously reverse both glasses. (That is, if a glass is brim down it is turned brim up, and vice versa). Do the same with another pair of glasses. Continue reversing pairs as long as you please. Is it possible to end with all six glasses upright? With all six upside down? Can you prove your answers mathematically?*

The concept of parity is applied to rotating figures in 3-space in the following manner. Consider the rotating cylinder drawn with solid black lines in Figure 58. Its structure can be described by reference to a coordinate system of three mutually perpendicular axes, traditionally labeled x, y, z as shown. The position of any point on the cylinder is given by an ordered set of three numbers. The first number is the

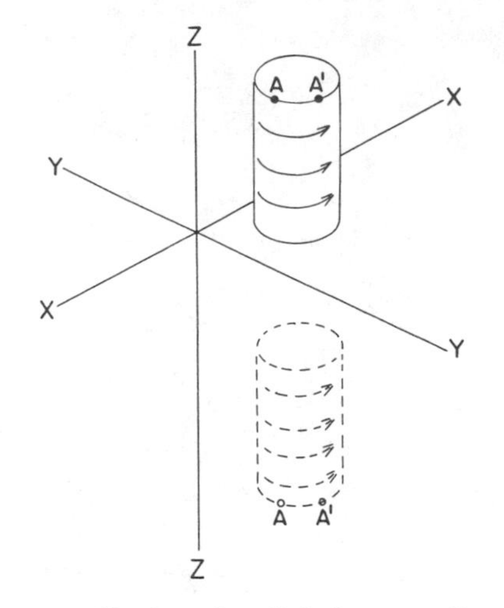

Figure 58. A rotating cylinder has even parity.

point's distance, measured along the x-axis, from a plane passing through the center of the coordinate system and perpendicular to the x-axis. The second number is the distance of the point, measured in similar fashion along the y-axis. The third number is the distance on the z-axis.

The cylinder drawn with dotted lines is the figure that results when all the z coordinate numbers, in the triples that designate the cylinder's points, have been changed in sign from plus to minus. Note that as the upper cylinder rotates in the direction of the arrows, point A on its upper edge moves from A to A'. The positions of A and A' on the dotted cylinder show that it is rotating in the same direction. True, the cylinder has been turned upside down by this transformation, but since the ends of the cylinder are indistinguishable, the upper and lower cylinders (including their spins) are superposable. In short, the entire system remains unchanged by the change of the sign for all z numbers.

Consider now the rotating cone drawn with solid lines in Figure 59. Below it is the cone that results when the z coordinate numbers are changed from plus to minus. Are the two figures superposable? No, they are mirror images of each other. If you turn the top cone upside down so that it coincides, point for point, with the bottom cone, then the spins will be in opposite directions. And if you turn the cones so

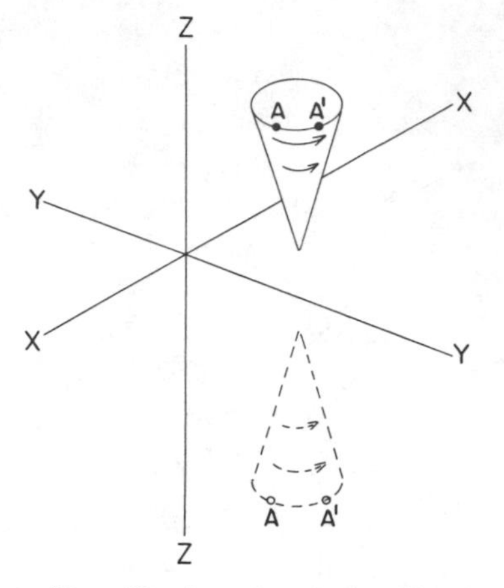

Figure 59. A rotating cone has odd parity.

that their spins coincide, the cones will point in opposite directions. The rotating cone is an asymmetric system possessing handedness.

It is not hard to see that any symmetric system in 3-space remains unchanged by a change in the sign of any one coordinate. Such systems are said to have an even (or definite) parity. Asymmetric systems are transformed to mirror images by a change in the sign of one coordinate. Such systems are said to have an odd (or indefinite) parity. The three coordinates, each of which can be plus or minus, behave in a manner somewhat like the three pennies, each of which can be heads or tails. If the system is asymmetrical, any *odd* number of sign changes has the same effect as changing one sign: it mirror-reflects the system. If you change the signs of all three axes, the system is reflected, because 3 is an odd number. Each single change produces a mirror reflection, but if a mirror reflection is reflected, you are back where you started. Every even number of sign changes leaves the system unaltered with respect to left and right. (This is why the two trick mirrors, described in chapter 3, give unreversed images; they reverse *two* axes of the coordinate system.) Every odd number of sign changes transforms it to its mirror image. Of course if the system is symmetrical (has even parity) then any number of sign changes, odd or even, leaves the system unchanged.

Physicists found it useful, in the 1920s, to apply these mathematical concepts to the wave functions that describe the elementary particles. Each function contains x, y, and z space coordinate numbers. If a change in the sign of one (or all three) coordinate numbers leaves the function unaltered, the function is said to have even parity. This is indicated by assigning to the function a quantum number of + 1. A function that is space-inverted (mirror-reflected) by a change in the sign of one (or all three) coordinate numbers is said to have odd parity. This is indicated by a quantum number of – 1.

Theoretical considerations (such as the left–right symmetry of space itself) as well as experiments with atomic and subatomic particles indicated that, in any isolated system, parity was always conserved. Suppose, for example, that a particle with even (+1) parity breaks down to two particles. The two new particles can both have even parity or both have odd parity. In either case the sum of the parities is even because an even number plus an even number is even, and an odd number plus an odd number is even. To say the same thing differently, the *product* of the two parity numbers is +1. (+1 times +1 is +1, and – 1 times – 1 is also + 1.) The final state of the system has a total parity of + 1. Parity is conserved. If an even particle should break down into two particles, one even and the other odd, the total parity of the final state would be odd. (An even number plus an odd number is an odd number, or +1 times – 1 is – 1.) Parity would not be conserved.

It is important to realize that we are no longer dealing with simple geometrical figures in 3-space but with complex, abstract formulas in quantum mechanics. It is impossible to go into more technical details about the exact meaning of parity conservation in quantum theory or the many ways in which it turned out to be a useful concept. Fortunately, the implications are not hard to understand. In 1927 Eugene P. Wigner was able to show that parity conservation rests squarely on the fact that all the forces involved in particle interactions are free of any left–right bias.[1] In other words, any violation of parity would be equivalent to a violation of mirror symmetry in the basic laws that describe the structure and interaction of particles. Physicists had long known that mirror symmetry prevails in the macroworld of whirling planets and colliding billiard balls. The conservation of parity suggests that this mirror symmetry extends down into atomic and subatomic levels. Nature, apparently, is completely ambidextrous.

This does not mean that asymmetry cannot turn up in the universe in all sorts of ways. It only means that anything nature does in a left-handed way she can do just as easily and efficiently in a right-handed way. For example, our sun moves through the galaxy in such a direction that the earth's motion with respect to the galaxy is along a helical path. Here is a clear instance of astronomical asymmetry. But this asymmetry is merely an accident in the evolution of the galaxy. Other planets, orbiting other suns, no doubt trace helical paths of opposite handedness. Our bodies have hearts on the left. Again, no fundamental asymmetry in natural law is involved. The location of the human heart is an accident in the evolution of life on this planet. In theory a person could be constructed with a heart on the right; in fact, as we have noted, such persons actually exist. Here we have an instance of an asymmetric structure that exists in both left and right forms, but one form is extremely rare. The parity conservation law does not say that mirror images of asymmetric structures or moving systems must exist in equal quantities. It merely asserts that there is nothing in nature's laws to prohibit the possible existence of both types of handedness.

Physicists sometimes explain the mirror symmetry of the universe in this way. Imagine a motion picture taken of any natural process. The film is mirror-reflected and projected on a screen so that you see a reversed movie of what actually occurred. Is it possible to examine this reflected motion picture and tell if it has been reversed? No, said the physicist in the 1940s, it is not. Of course, we could recognize at once that it was reversed if we saw in the film any man-made asymmetric structures, such as printed letters or numbers, the face of a clock, and so on. But we are concerned only with the fundamental processes of nature, uncontaminated by the artificial asymmetry introduced by living things. Perhaps we are watching drops of oil falling into water, or a chemical reaction taking place. There is no way, physicists said in the 1940s, that we can tell if such a film has been reversed.

If we took a motion picture showing the growth of left-handed crystals from a left-handed compound, it is true that a reversal of the film would show right-handed crystals being formed. But unless we had advance information, we would have no way of knowing that we were not watching an unreflected motion picture of the growth of right-handed crystals from a right-handed compound. Suppose we paint the north end of a magnetic needle red, then take a color motion picture of the needle-and-wire experiment that shocked Mach. The reversed picture would, it is true, show the red end pointing the wrong way. But if we saw such a picture without having previously been told how it was made, we could assume that someone had painted the *south* end of the needle red and all would be well. If magnets do not have their poles labeled *N* and *S*, or distinguished in some other way, a reflected picture of an experiment involving them does not provide any clue by which one can be sure the film has been reversed.

All this is, of course, just another way of stating the Ozma problem. If an experiment could be performed that violated the law of parity, that showed a basic preference of nature for either right or left, we would immediately have a solution to the Ozma problem. We would simply explain to the scientists of Planet X how to set up such an experiment. From its asymmetric twist we and they could easily arrive at a common understanding of left and right.

I once wrote a short-short science-fiction story called "Left or Right?" (*Esquire*, February 1951) wherein the law of parity was dramatized as follows. The earth has been caught off-guard by a sneak attack by the members of Planetary System Zeta-59. A factory in Alaska, which manufactures small helical devices called helixons, has been demolished. Helixons are essential for the earth's defense system. Now there is a fatal shortage. The nearest source of a new supply is a planet halfway across the galaxy—a planet that was colonized centuries ago by terrestrials. A group of astronauts is sent on the mission of bringing back from this planet a new supply of helixons. On the way back, with a shipload of helixons, the ship is struck by a meteor that flips it over several times in 4-space before dropping it back into 3-space. The ship lands on an unknown planet to make repairs. The planet is in an uncharted part of the galaxy.

It suddenly occurs to the ship's captain that if the ship made an odd number of somersaults in 4-space, the entire ship has been mirror-reflected. All the helixons will have changed handedness and therefore be useless. How can he determine, before he completes his trip home, whether this has occurred or not? Studying asymmetric structures on the ship—the words on charts, for instance—is no help, because if the ship is reversed, so are the astronauts. The printing would seem normal because the sides of their brains would have been exchanged.

The planet is not inhabited, but of course its substances and natural laws are the same as on earth. The spaceship has a well-equipped laboratory. Can an experiment be performed that will indicate whether the ship has been reversed? The captain realizes that there is no such experiment. The laws of nature are mirror-symmetric. Parity is conserved. Even if he found carbon life in some form on the planet—organisms containing asymmetric amino acids—it would be of no help. There are no laws of nature to prevent the amino acids from being right-handed.

Six years after I wrote this story, it became hopelessly obsolete. In 1957 the law of the conservation of parity was overthrown. At Columbia University an experiment was performed in which a symmetrical nuclear system changed to an asymmetric one. A basic handedness was revealed in the laws that describe the structure of certain elementary particles when they undergo a certain type of reaction. If my perplexed space captain had had the necessary equipment, he could have performed this experiment on the planet where he had landed; from it he could have deduced whether his ship had been reversed. The Columbia experiment is not reflectible. A motion picture of it, projected with reversed film, is a picture of an experiment that could not be performed anywhere in the galaxy. It is an experiment that solves the Ozma problem.

Chapter 22 will detail the story of this "gay and wonderful discovery" (as J. Robert Oppenheimer called it) and trace some of its revolutionary implications. But first we must take a close look at what physicists call the antiparticles, and at a strange, hypothetical variety of matter known as antimatter. Antiparticles are intimately bound up with the overthrow of parity. Knowing something about them will make the story of parity's downfall much easier to understand.

Notes

1. In 1963 Wigner received a share of the Nobel Prize in physics for his pioneer work on the symmetry principles underlying particle interactions. To laymen he is best known as one of the signers of Einstein's famous letter to President Roosevelt, informing him of the possibility of an atomic bomb, and for having suddenly produced, as if by magic, a

bottle of Chianti to celebrate the occasion, in 1942, when Fermi and his associates achieved the first self-sustaining chain reaction.

I once had the pleasure of asking Wigner how he had performed this surprising trick. He replied by asking if I knew Poe's story, "The Purloined Letter." As in the story, Wigner's method of concealment was to put the bottle in something so commonplace and obvious that it would not be noticed. He had carried the bottle to the site in an ordinary paper bag, which he simply placed aside until the right moment.

21. ANTIPARTICLES

The history of theories of matter has swung like a pendulum from simplicity to complexity to simplicity to complexity. The first swing was long and slow. The Greeks had a simple theory in which all substances were combinations of four elementary types of matter: earth, air, fire, and water. It was not until two thousand years later that the facts of chemistry made it necessary to recognize about eighty different elements, each composed of its own individual type of atom. These atoms were the "elementary particles" until the beginning of the present century, when the pendulum began a fast swing back to simplicity. By the early 1930s the differences between atoms could be explained elegantly by assuming only *three* (one less than Aristotle's four) elementary particles: protons, neutrons, and electrons.

Then the pendulum swung rapidly again. Today, physicists have identified hundreds of elementary particles, the count depending on which are called elementary and which are considered different states of the same particle. This newly discovered complexity annoys physicists as much as the complexity of the periodic table of elements annoyed them before they learned how to "explain" the table by the Bohr model of the atom and its later refinements. The new particles are, in C. P. Snow's excellent metaphor, as "grotesque as a stamp collection." They have, said J. Robert Oppenheimer, an "insulting lack of meaning."

No one knows when the pendulum will swing back to simplicity. Some particle physicists think that before many years have passed an elegant new theory, based on a few simple mathematical assumptions, will explain why the particles are just what they are and not something else. An enormous step in this direction was the discovery by Murray Gell-Mann (and independently by Yuval Ne'eman) of a striking pattern of classification called the eightfold way (after a Buddhist religious phrase) because it involves assigning eight quantum numbers to each particle for eight different quantities that are conserved. The quantum numbers are then related to one another by the symmetries of a simple group structure known to mathematicians as a Lie group (after Marius Sophus Lie, a Norwegian mathematician). This eightfold pattern was strongly confirmed by the finding of a new particle called the omega minus. Many peculiar properties of the new particle had been predicted by the pattern: a truly remarkable instance of the power of group theory, which had been introduced into quantum mechanics by Wigner, to predict the properties of new particles. In Snow's metaphor, the eightfold way is a way of pasting seemingly random stamps on the page of a scrapbook to form a beautifully symmetric design of shapes and colors. Particles are less "insulting" when they can be elegantly classified.

Physicists do not agree on why the eightfold way is such an accurate classification. New theories are constantly being put forth. At the moment the two leading contenders are the "bootstrap" (or "democratic") theory, created and eloquently championed by Geoffrey Chew, and the quark theory of Murray Gell-Mann and George Zweig.

The bootstrap hypothesis maintains that there *are* no fundamental particles; put differently, all are equally basic. Like a man supporting himself in midair by lifting on his bootstraps, the particles are a single family in which each member is held up by a combination of the others. Each particle is a composite interaction of other particles. It is as if *A* is made of *B* and *C*, *B* is made of *A* and *C*, and *C* of *A* and *B*.

The quark theory asserts that all particles are combinations of more elementary units that Gell-Mann named quarks after a line in James Joyce's *Finnegans Wake*: "Three quarks for Muster Mark!" At first it was believed that there are only three varieties: up, down, and sideways (or "strange"), together with their antiparticles. The three basic types are called flavors. There are now excellent reasons for thinking there is a fourth flavor called charm, a new quantum number proposed by Sheldon Lee Glashow. Charm demands a new type of quark—the charmed quark. Each flavor comes in three colors (in the United States the colors are red, white, and blue). This makes twelve quarks in all, with their twelve antiquarks. Properties such as color and charm are, of course, just whimsical names, with almost no relation to their usual meaning. Some theorists think there may be still other quarks, with new quantum numbers such as truth, beauty, and sex.

The evidence for the charmed quark got a big boost in 1974 when the first of a new family of particles were found that can best be explained as a brief mating of a charmed quark with its charmed antiquark. (Since the charms cancel, the union is not at all charming). These particles are extremely short-lived, lasting about one ten-thousandth of a billionth of a second. In a few pages we will see how the evanescent mating of electron and positron make up a pair called positronium. Some physicists call the new union of charmed quark with its antiparticle charmonium. Others have suggested replacing *charm* with *panda* (after the panda's renowned shyness) and calling the new substance pandemonium.

The first particle of pandemonium was found simultaneously in this country's West, at Stanford University, and in the East at the Brookhaven National Laboratory in New York. The West group called it J, the East group, psi. As an East–West compromise, Richard P. Feynman proposed calling the family pions, using J for the first member to be discovered.

As for the quarks themselves, physicists are still searching for them without finding any. It may be the nature of the beasts that they cannot be isolated from particles to which they are attached. A quark and its antiquark have been compared to the ends of a piece of string. The string is real, and so are its ends, but you can no more separate a pure end from a string than you can separate a single pole from a magnet. Other theorists think that quarks are even less real than string ends. They are merely fictitious animals invented to make the equations come out right.

Quarks may prove to be identical with another set of mathematical constructs by Richard Feynman. He calls them partons. Abdus Salam is promoting a "quark liberation movement," which regards quarks as composite entities made up of still smaller things called prequarks, or preons. In Peking a group of physicists have proposed a similar view involving stratons, perhaps infinitely nested like a set of Chinese boxes.

The astonishing thing about the two main theories, the bootstrap and quark, is how each both reinforces and contradicts the other. Both predict exactly the same new particles! Fritjof Capra, in an interesting book called *The Tao of Physics* (1975), likens the quark conjecture to a Zen koan (a statement intended to provoke insight by its absurdity) and the bootstrap theory to the Eastern view of the unity of all things. Robert H. March, discussing the two views in *Science Year 1973* (Field Enterprises, 1972), puts it this way: "The implications are uncanny. If the quark theory is correct, the particles formed from quarks are engaged in a vast conspiracy to make the bootstrap look good. If the bootstrap is right, the only way for nature to be consistent is to behave as if quarks existed, even though the bootstrap theory rules them out."

Some physicists expect that this curious state of affairs will soon evaporate when a genius comes along with the right flash of insight that explains both theories with a deeper theory. At the moment, the most exciting work in theoretical physics is the work on gauge theories. These are field theories concerning fundamental laws that vary from point to point in space-time but preserve a basic symmetry. The goal is to unite the strong, weak, and electromagnetic forces—perhaps even gravity—in one fundamental theory. One of the leaders in this work is Steven Weinberg, whose recent book *The First Three Minutes* (Basic Books, 1977)—a reconstruction of what may have happened in the first three minutes after the big bang—I highly recommend.

Other physicists are not as optimistic as Weinberg about finding a simple, elegant theory that will explain all the particles and the forces that govern their interactions. They anticipate a slowing down of the pendulum and suspect that no completely adequate theory can be formulated until many more facts are in; facts they fear will be hard to come by. Even if the eightfold way turns out to be as accurate a classification as the periodic table, it may be decades until the pattern itself can be explained adequately by fundamental laws.

Will the pendulum ever stop swinging? Are there endless levels of microstructure? "It has not been necessary," wrote Edward Teller in

Our Nuclear Future (Criterion, 1962), "to ascribe an internal structure to the electron." To this sentence he appended the following footnote: "Yet."

That trim little trio—proton, neutron, electron—was not firmly established until 1932, when James Chadwick, at the Cavendish Laboratory in Cambridge, finally trapped the neutron.[1] Its existence had been long suspected, and physicists heaved a great sigh of relief when it was finally identified. Before the year ended, however, their complacency received a rude jolt. Carl David Anderson, at the California Institute of Technology, was examining some cloud-chamber photographs of cosmic-ray tracks when he came upon the path of a particle that should have been an electron except that the path curved the wrong way. After considering and rejecting various explanations of this anomaly, Anderson finally concluded that the track could have been made only by an electron with a positive charge. He dubbed it a *positron* and the name has stuck.

The positron was the first antiparticle to be discovered. Every elementary particle is now known to have a corresponding antiparticle. The two particles are alike in all respects except that they are opposite in the sign of any quantity (represented by a plus or minus quantum number) that is conserved. If a particle is charged, its antiparticle has an equal but opposite charge. If it has a magnetic moment, its antiparticle has a magnetic moment of opposite sign. The *K*-meson and anti-*K*-meson have neither charge nor magnetic moment but are opposite in the values of the quantum number called strangeness. In other words, all conserved quantities must be of opposite sign so that when the particle and antiparticle come together, these quantities will cancel each other, leaving nothing but pure energy (photons). In the case of the photon and the neutral pi meson, particle and antiparticle are one and the same.

Before Anderson made his discovery, most physicists had been reluctant to admit that antiparticles could exist. There was one notable exception. Paul Adrien Maurice Dirac, one of the most creative mathematical physicists of all time, had proposed a "hole" theory of particles, which predicted the existence of antiparticles. Dirac's theory is impossible to make clear without high-order mathematics, but we can get a crude (very crude) idea of it by considering Sam Loyd's 15-puzzle. The object is to slide the little squares around by continually pushing a square into a vacant hole, and in this way obtain various patterns of numbers.[2] Just as the little squares move about by discrete "quantum jumps" from one position to an adjacent one, so does the "hole." It, too, travels from position to adjacent position, behaving mathematically exactly like one of the squares. In fact, the theory of the 15-puzzle is usually explained by treating the hole as a "thing" that moves about within the frame.

Dirac's theory resembles the 15-puzzle in the following way. It assumes that empty space is not really empty: it is a vast, compact sea of particles, all possessing negative inertial mass. (This means that if a force acts on such a particle it moves it in a direction *opposite* to that in which the force is acting.) Under certain conditions a particle can be dislodged from the sea and raised, so to speak, to a level outside the sea. When this occurs, there is a simultaneous "pair creation" of two types of electrons, both with positive inertial mass. One is the ordinary electron, with its negative charge; the other is the "hole" left behind in the sea. The hole is a "thing" in the same sense that a moving bubble in a liquid is a thing, or the moving hole in Sam Loyd's puzzle.[3] In Dirac's theory it would behave like an electron with a positive charge. It would be, Dirac wrote in 1931, "a new kind of particle, unknown to experimental physics, having the same mass and opposite charge to an electron. We may call such a particle an anti-electron."

The antielectron, Dirac continued, would not last very long in this world. For an instant it would "move" (as other particles in the sea shifted around), then an electron would fall into the hole and there would be simultaneous pair destruction. The two particles would annihilate each other and vanish from our observation. In similar fashion, Dirac reasoned, protons would have their own sea of densely packed particles. Under certain conditions a particle would be knocked out of this sea to become an ordinary proton, leaving a hole of negative charge that would behave like an antiproton.

All this in 1931! Did Anderson know of Dirac's remarkable theory? No, he did not. In fact, when he looked it up and read it, after his discovery of the positron, he confessed that he could not fully understand it. So, in his own way, Anderson had as much insight and courage as did Dirac. Without any theoretical justification, he stared at that puzzling track on his famous cloud-chamber photograph and concluded that the evidence could not be explained away by traditional theory. It was the unmistakable track of a positive electron.

Other physicists lost no time in confirming Anderson's discovery. In a few months, at many laboratories, atomic nuclei were bombarded with gamma rays to produce pairs of electrons and positrons. As Dirac had predicted, the positron was short-lived. As soon as it met an electron (and there were plenty of electrons around to meet), there was mutual pair annihilation. Later it was discovered that, just before the two particles destroy each other, they spin around a common center, creating for a fleeting instant an atom of what physicists call positronium. A quick dance of death—then poof! The two particles vanish, sending off gamma rays of two or three photons, the number depending on whether the pair danced with their magnetic axes parallel (north poles pointing the same way) or antiparallel (north poles pointing in opposite directions).

Dirac's theory, as we have seen, also predicted an antiproton. It could be created only in combination with a proton and would be annihilated as soon as it encountered another proton. It was not until 1955, twenty-three years after Anderson's discovery of the antielectron, that a group of physicists at the University of California, Berkeley, using a powerful accelerator called the Bevatronn succeeded in creating the first proton-antiproton pair.[4] The couple behaved just as Dirac had said they would.

A year later, in 1956, Berkeley scientists working with the Bevatron identified for the first time an antineutron. Though the neutron has no electrical charge, it does possess spin and a magnetic moment. How it can have a magnetic field without having an electrical charge is still a mystery, because magnetic fields are generated only by moving charges. There are several theories to account for it. For example, the neutron may be a complicated structure in which a negatively charged particle circulates around a positively charged core, which may be an ordinary proton. The two charges cancel each other, but the moving negative charge induces a magnetic field. Recent experiments have cast doubt on this theory. A better view is that the neutron's core is uncharged but around it orbit an equal number of positive and negative particles. If the positive particles whirled one way and the negative the other, their magnetic axes would all point in the same direction to create an overall magnetic field.

In any case, the neutron does have a magnetic field, and it is the reversal of the axis of this field that identifies the antineutron. When a proton and antiproton have a near miss, instead of destroying each other they neutralize each other's charge. The proton turns into a neutron and the antiproton turns into an antineutron. Something neat and delightful is obviously taking place here, although no one knows yet exactly what.

Since 1956 physicists have found that every elementary particle, with the two exceptions previously noted (the photon and the neutral pi meson), has its antiparticle twin. As soon as it became apparent that the three particles of ordinary matter (proton, neutron, electron) have antiparticles, physicists said to themselves: Why not antimatter? An atom of antihydrogen would have an antiproton nucleus around which would whirl a positron (antielectron) with a positive charge. Antideuterium, the simplest isotope of antihydrogen, would have a similar structure except that the antinucleus would also contain an antineutron. Similarly for all the other elements. Each antiatom would be exactly like an atom except that it would be constructed of antiparticles instead of particles. There is no reason why antiatoms could not link themselves into antimolecules to form antielements and anticompounds that would be exact counterparts of those we know. Antiwater would be formed by the union of two antiatoms of antihydrogen and one antiatom of antioxygen.

At the time this is written, not a single antiatom of antimatter has been discovered or created in the laboratory,[5] but physicists see no theoretical reason why such matter could not exist. Of course, the instant a bit of antimatter came in contact with matter, there would be a bang. In fact, the explosion would be much more powerful than the explosions of atomic or hydrogen bombs. In these bomb explosions only part of the mass of whatever substance is involved is converted to energy. If matter combined with antimatter, virtually *all* the mass would become energy. First it would produce pi mesons and other particles; then these particles would decay immediately into neutrinos and radiation leaving the scene with the speed of light. It would be the ultimate explosion.

Science has not yet found a way to blow the entire earth to smithereens. It would be easy to destroy all life on the planet (by a variety of techniques), but the power to disintegrate the earth itself has not yet been discovered. Such power might be available if antimatter could be created in large enough amounts. (To prevent immediate explosion it would have to be kept suspended in a vacuum, isolated from matter). Are the asteroids, those myriads of rocky hunks that circle the sun between the orbits of Mars and Jupiter, the remnants of a planet whose scientists finally discovered how to make antimatter? Perhaps it is part of God's vast cosmic plan to allow life to evolve on millions of planets in the hope that somewhere in the universe intelligent creatures may develop who are capable of discovering the secrets of matter without blowing themselves into eternity. The planet just beyond Mars failed to make the grade. The earth is now on the Brink of the Great Test.

Of course, all this is old stuff to science-fiction fans. As soon as the physicists predicted antimatter, writers of science fiction began to play with the idea. (At first they called it "contraterrene" matter, but the term is now obsolete.) Boy meets antigirl; they kiss; the end. James Blish's novel *The Triumph of Time* (Avon, 1958) is woven around antimatter themes. It is evident that our galaxy must consist entirely of matter, but separated from our galaxy by inconceivably vast distances are other galaxies. Are some of them made of antimatter? There is no way to tell from the light they send us, because the light quantum, the photon, is identical with its antiparticle. Any antiparticles shot out from an antigalaxy would be annihilated long before they came near the earth (except possibly for the antineutrinos that we will meet in chapter 23.)

In the constellation of Cygnus (the Swan) are two galaxies which seem to be passing through each other and sending out radio energy much greater than can be accounted for. Some astronomers have wondered if we are seeing here a collision of galaxy and antigalaxy. Other astronomers think not. It has been suggested that meteors of antimatter may occasionally strike the earth, such as the mysterious object that crashed in Siberia on June 30, 1908, causing a monstrous explosion but leaving no trace of meteoric fragments. This seems unlikely. All meteors are believed to come from our galaxy and would therefore be made of ordinary matter.

The possibility of creating small quantities of antimatter to use as fuel for spaceships is taken quite seriously by physicists, though at present no one has any notion of how to go about making it. It would be, of course, a kind of ultimate fuel. Presumably antiiron could be magnetized and kept suspended in a vacuum by magnetic fields, then by some ingenious method made to combine slowly with ordinary iron.

In 1956 the *San Francisco Chronicle* reported a speech by Edward Teller in which the famous physicist had discussed antimatter and the fact that it would explode on contact with ordinary matter. This inspired physicist Harold P. Furth, at the University of California's Lawrence Radiation Laboratory, to write a poem: "Perils of Modern Living." The *New Yorker* printed it on page 52 of the November 10, 1956, issue:

Well up beyond the tropostrata
There is a region stark and stellar
Where, on a streak of anti-matter,
Lived Dr. Edward Anti-Teller.

Remote from Fusion's origin,
He lived unguessed and unawares
With all his anti-kith and kin
And kept macassars on his chairs.

One morning, idling by the sea,
He spied a tin of monstrous girth
That bore three letters: A.E.C.
Out stepped a visitor from Earth.

Then, shouting gladly o'er the sands,
Met two who in their alien ways
Were like lentils. Their right hands
Clasped, and the rest was gamma rays.

EXERCISE 15: *As we will learn in chapter 23, antimatter is now believed to involve, in addition to the reversal of charges and magnetic axes, a reversal of left and right. Assuming that Teller and Anti-Teller are exact enantiomorphs, describe the possible interpretations that can be given to the phrase: "Their right hands clasped."*

Dr. Teller's response to the poem was an amusing letter that appeared in the *New Yorker*, December 15, 1956:

University of California
Radiation Laboratory
Berkeley, California
November 26, 1956

To the Editors, *The New Yorker*,

Dear Sirs:

In a recent issue of *The New Yorker*, I found the following poem, describing the meeting of Dr. Edward Anti-Teller with an imagined person differing from Anti-Teller only in the sign of the charges carried by the particles in his body. [The poem is reprinted.]

The meeting, as described, is interesting, and tempts me to offer some scientific details.

I do not believe that Anti-Teller lives in our galaxy, since it is unlikely that there are any anti-stars or anti-planets in our milky-way system. On the other hand, anti-galaxies may exist. The main questions are how to get there and what to expect upon arrival. (I shall not worry about the mechanics of space travel. Every child knows that it is feasible.)

The distance is somewhat of an obstacle. Light takes more than a million years to travel to the next spiral nebula. Fortunately, Einstein has shown that a million years will seem like only a few years if one travels fast enough, and so an explorer might arrive during his lifetime, though not during the lifetime of his friends whom he left behind on Earth. As he approaches the anti-galaxy, he will be attracted by anti-gravity. In fact, gravity and anti-gravity are one and the same thing. Here some may disagree, but upon second thought they will find they are wrong.

As the traveller enters the anti-galaxy, his ship will be bombarded by anti-particles. This bombardment will heat the space ship. He must not crowd the speed limit (which is the speed of light), or his ship will melt. Furthermore, the resulting radiation will kill him before he has penetrated as much as a millionth of the anti-galaxy. But let us not give up; Anti-Teller may live near the edge of the anti-galaxy.

At a distance of about two hundred miles from the surface of

Anti-Earth, the intruder will surely be killed by the annihilation radiation that is produced as the space ship begins to dip into the anti-atmosphere. Only a miracle or an unexpected development in biophysics can save him. Before he gets down to an altitude of a hundred miles, the space ship will collapse and *nothing* can save him.

But let us arrange a meeting between Teller and Anti-Teller on a truly neutral ground: in space. If they are appropriately dressed (space suit and anti-space suit, respectively), and if they carefully avoid the escape of any molecules or anti-molecules, they may approach without danger. They can see each other without trouble, because light and anti-light are the same. Upon contact, however, a violent explosion will occur. Parts of Teller and Anti-Teller will produce an assortment of ephemeral particles (known as mesons, hyperons, and anti-hyperons) and a great number of more stable products, such as nuclear fragments, anti-nuclear fragments, electrons, positrons, neutrinos, anti-neutrinos, and gamma rays. The remainder will fly apart in opposite directions as vapor and anti-vapor. All this will happen faster than anti-thought, which is probably the same as thought.

In spite of this inauspicious prospect, I was pleased that *The New Yorker* mentioned me. Come to think of it, only Anti-Teller was mentioned by name in the poem, but I am confident that somewhere in an anti-galaxy *The Anti-New Yorker* devoted some pleasant lines to

Yours sincerely,
EDWARD TELLER

It is important to understand that the discovery of antiparticles did not in any way violate the law of parity. As we have seen, the distinction between north and south poles of magnetic fields does not lead to any solution of the Ozma problem—that is, it does not indicate any fundamental preference of nature for left or right. Similarly, no left–right bias is indicated by the distinction between positive and negative charge. Like north and south pole, positive and negative charge are merely conventional labels for two opposite states of electrical energy. Magnetic force is now understood in the sense that it is reduced to a force field created by the motion of an electric charge, and we have seen how the direction of spin of such charges explains the difference between the two ends of a magnetic axis. Why electrical energy should divide into the two states of positive and negative is still a total mystery. Physicists simply accept it as one of the given facts of existence.

The two charges are distinguished from each other by the fact that opposite charges attract each other, like charges repel. Every known particle has either a negative electrical charge of one quantum, a positive charge of one quantum, or no charge at all. (In quantum mechanics the charge is expressed by the quantum numbers: + 1, – 1, and 0.) Exactly what these labels stand for no one knows. The point to be emphasized here is that this labeling describes a state of affairs that does not in any known way involve a violation of left–right symmetry.

However, when the electric charges and magnetic axes are both taken into account, we can diagram a particle and its antiparticle in such a way that each appears to be a mirror image of the other. For example, the diagrams of an electron and positron are shown in Figure 60. A proton and antiproton diagram are shown in Figure 61. These are no more than symbolic models of a state of affairs that can be expressed accurately only by the wave functions of quantum mechanics. Nevertheless, like the diagrams of molecules in which atoms are shown joined by chemical bonds, such schematic drawings are enormously useful and often suggestive of theoretical possibilities.

In looking at these diagrams the thought that immediately occurs is: Perhaps the antiparticle really *is* a mirror reflection of a particle. The only difference between the right and left particles in each picture, aside from the mirror reflection of their structure, is that one has a positive charge, the other a negative charge. Could it be that the distinction between positive and negative charge may rest, in some presently unknown fashion, on some sort of asymmetric spatial structure in the particle itself? Will future investigations of the electron's structure (not "yet" possible, as Teller's footnote reminds us) disclose some type of true spatial asymmetry, just as the investigation of chemists in the last century disclosed that Pasteur's optical isomers were true mirror images of each other? Remember how van't Hoff's colleague contemptuously dismissed his speculations along such lines as "miserable speculative philosophy"?

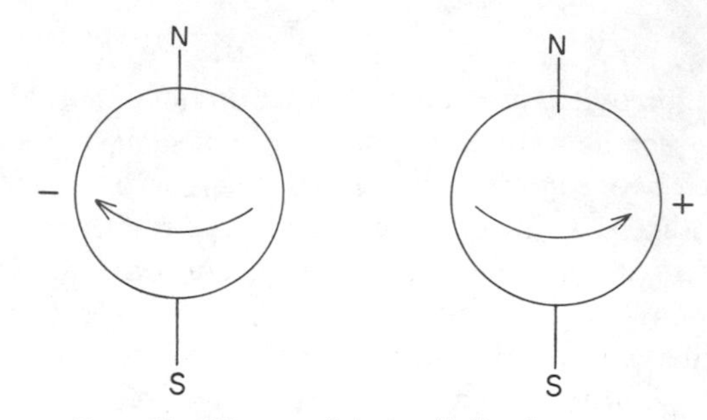

Figure 60. Diagrams of electron (left) and positron (right).

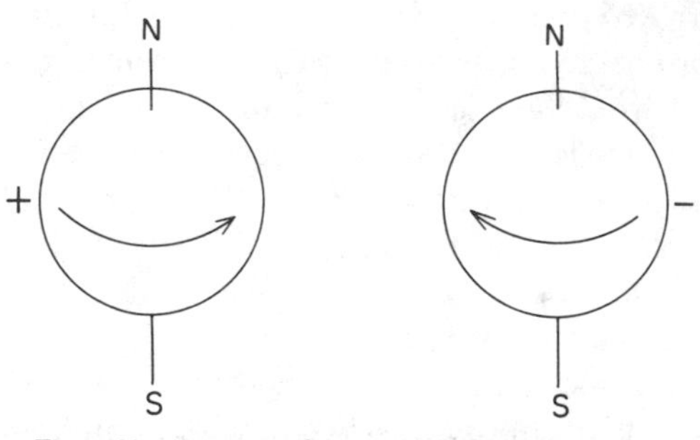

Figure 61. Diagrams of proton (left) antiproton (right).

Pasteur and van't Hoff had strong intuitive hunches that rested on the same insight as Kant's puzzlement about his ears. How can two things be exactly alike in all respects and yet somehow be different? In an analogous way, how can the electron and positron be exactly alike in all respects yet differ in electric charge? The mirror-reflection diagrams suggest one possible answer: they really *are* alike except that one of them, somehow, "goes the other way."

Even after the discovery of antiparticles, physicists did not seriously entertain the thought that an antiparticle might be a true mirror reflection of an unknown asymmetric structure. The reason was simple: If there were a spatial asymmetry of any sort in the structure of particles, it would surely manifest itself in some kind of violation of parity. That is to say, there should be an experiment in which the asymmetry of a particle should lead to some kind of *measurable* (not just symbolic or diagrammatic) spatial asymmetry, a measurable left–right bias, a measurable handedness. No such experiment was known. Parity was always conserved.

Then in the years from 1954 to 1956 a curious situation developed with respect to two particles called the theta meson and the tau meson. How this puzzle, as it was called, led to the overthrow of parity is an exciting story that will be told in the next chapter.

Notes

1. It was also at Cambridge that James J. Thomson, in the late 1890s, discovered the electron. The proton's existence was firmly established about fifteen years later by Ernest Rutherford (later Lord Rutherford), then at the University of Manchester.

2. The patterns of the 15-puzzle fall into two mutually exclusive sets of opposite parity. Parity is preserved in the sense that once the squares are placed in the frame to form an initial pattern, only patterns of the same parity can be obtained by sliding the squares. For a brief discussion of the theory of the puzzle and its connection with odd and even numbers see my *Scientific American Book of Mathematical Puzzles and Diversions* (Simon and Schuster, 1959), pp. 86–89.

3. The concept of particles of matter as bubbles of nothing in a sea of particles is older than Dirac's theory. It was used by the Irish physicist Osborne Reynolds in his granular theory of the universe (see his *On an Inversion of Ideas as to the Structure of the Universe*, 1902, and *The Sub-Mechanics of the Universe,* 1903, both issued by Cambridge University Press). It is implied in the earlier "ether squirt" theory, of the English scientist Karl Pearson, in which particles are regarded as points at which ether is squirted into 3-space from 4-space (see *The American Journal of Mathematics,* vol. 13, 1891, pp. 309–62).

4. In 1959 Emilio Gino Segrè and Owen Chamberlain received the Nobel Prize in physics for their work in this first demonstration of the existence of the antiproton.

5. The antiproton, the nucleus of antihydrogen, has been observed, but not with a positron around it in stable orbit to make an atom of

antihydrogen. The nucleus of antideuterium (see chapter 19), or heavy hydrogen, has also been observed. Russian physicists have reported observing the nucleus of antitritium (tritium is another isotope of hydrogen), and the nucleus of an isotope of antihelium. But no stable atom of antimatter has yet been observed.

22. THE FALL OF PARITY

As far as anyone knows at present, all events that take place in the universe are governed by four fundamental types of forces (physicists prefer to say *interactions* instead of *forces*, but there is no harm in using here the more common term):

1. Nuclear force.
2. Electromagnetic force.
3. Weak interaction force.
4. Gravitational force.

The forces are listed in decreasing order of strength. The strongest, nuclear force, is the force that holds together the protons and neutrons in the nucleus of an atom. It is often called the "binding energy" of the nucleus. Electromagnetism is the force that binds electrons to the nucleus, atoms into molecules, molecules into liquids and solids. Gravity, as we all know, is the force with which one mass attracts another mass; it is the force chiefly responsible for binding together the substances that make up the earth. Gravitational force is so weak that unless a mass is enormously large it is extremely difficult to measure. On the level of the elementary particles its influence is negligible.

The remaining force, the force involved in "weak interactions," is the force about which the least is known. That such a force must exist is indicated by the fact that in certain decay interactions involving particles (such as beta-decay, in which electrons or positrons are shot out from radioactive nuclei), the speed of the reaction is much slower than it would be if either nuclear or electromagnetic forces were responsible. By "slow" is meant a reaction of, say, one ten-billionth of a second. To a nuclear physicist this is an exceedingly lazy effect —about a ten-trillionth the speed of reactions in which nuclear force is involved. To explain this lethargy it has been necessary to assume a force weaker than electromagnetism but stronger than the extremely weak force of gravity.

The "theta-tau puzzle," over which physicists scratched their heads in 1956, arose in connection with a weak interaction involving a "strange particle" called the *K*-meson. (Strange particles were called "strange" because they did not seem to fit in anywhere with any of the other particles.) There appeared to be two distinct types of *K*-mesons. One, called the theta meson, decayed into two pi mesons. The other, called the tau meson, decayed into three pi mesons. Nevertheless, the two types of *K*-mesons seemed to be indistinguishable from each other. They had precisely the same mass, same charge, same lifetime. Physicists would have liked to say that there was only one *K*-meson; sometimes it decayed into two, sometimes into three pi mesons. Why didn't they? Because it would have meant that parity was not conserved. The theta meson had even parity. A pi meson has odd parity. Two pi mesons have a total parity that is even, so parity is conserved in the decay of the theta meson. But *three* pi mesons have a total parity that is odd.

Physicists faced a perplexing dilemma with the following horns:

1. They could assume that the two *K*-mesons, even though indistinguishable in properties, were really two different particles: the theta meson with even parity, the tau meson with odd parity.

2. They could assume that in one of the decay reactions parity was not conserved.

To most physicists in 1956 the second horn was almost unthinkable. As we saw in chapter 20, it would have meant admitting that the left–right symmetry of nature was being violated; that nature was showing a bias for one type of handedness. The conservation of parity had been well established in all "strong" interactions (that is, in the nuclear and electromagnetic interactions). It had been a fruitful concept in quantum mechanics for thirty years.

In April 1956, during a conference on nuclear physics at the University of Rochester in New York, there was a spirited discussion of the theta-tau puzzle. Richard P. Feynman[1] raised the question: Is the law of parity sometimes violated? In corresponding with Feynman, I received some of the details behind this historic question. They are worth putting on record.

The question had been suggested to Feynman the night before by Martin Block, an experimental physicist with whom Feynman was sharing a hotel room. The answer to the theta-tau puzzle, said Block, might be very simple. Perhaps the lovely law of parity does not always hold. Feynman responded by pointing out that if this were true, there would be a way to distinguish left from right. It would be surprising, Feynman said, but he could think of no way such a notion conflicted with known experimental results. He promised Block he would raise the question at next day's meeting to see if anyone could find anything wrong with the idea. This he did, prefacing his remarks with, "I am asking this question for Martin Block." He regarded the notion as such an interesting one that, if it turned out to be true, he wanted Block to get credit for it.

Chen Ning Yang and his friend Tsung Dao Lee, two young and brilliant Chinese-born physicists, were present at the meeting. One of them gave a lengthy reply to Feynman's question.

"What did he say?" Block asked Feynman later.

"I don't know," replied Feynman. "I couldn't understand it."

"People teased me later," writes Feynman, "and said my prefacing remark about Martin Block was made because I was afraid to be associated with such a wild idea. I thought the idea unlikely but possible, and a very exciting possibility. Some months later an experimenter, Norman Ramsey, asked me if I believed it worthwhile for him to do an experiment to test whether parity is violated in beta decay. I said definitely yes, for although I felt sure that parity would *not* be violated, there was a possibility it would be, and it was important to find out. 'Would you bet a hundred dollars against a dollar that parity is not violated?' he asked. 'No. But fifty dollars I will.' 'That's good enough for me. I'll take your bet and do the experiment.' Unfortunately, Ramsey didn't find time to do it then, but my fifty dollar check may have compensated him slightly for a lost opportunity."

During the summer of 1956 Lee and Yang thought some more about the matter. Early in May, when they were sitting in the White Rose Cafe near the corner of Broadway and 125th Street, in the vicinity of Columbia University, it suddenly struck them that it might be profitable to make a careful study of all known experiments involving weak interactions. For several weeks they did this. To their astonishment they found that although the evidence for conservation of parity was strong in all strong interactions, there was no evidence at all for it in the weak. Moreover, they thought of several definitive tests, involving weak interactions, which would settle the question one way or the other. The outcome of this work was their now-classic paper "Question of Parity Conservation in Weak Interactions."

"To decide unequivocally whether parity is conserved in weak interactions," they declared, "one must perform an experiment to determine whether weak interactions differentiate the right from the left. Some such possible experiments will be discussed."

Publication of this paper in *The Physical Review* (October 1, 1956) aroused only mild interest among nuclear physicists. It seemed so unlikely that parity would be violated that most physicists took the attitude: Let someone else make the tests. Freeman J. Dyson, a physicist now at the Institute for Advanced Study in Princeton, writing on "Innovation in Physics" (*Scientific American*, September 1958) had these honest words to say about what he called the "blindness" of most of his colleagues: "A copy of it [the Lee and Yang paper] was sent to me and I read it. I read it twice. I said, 'This is very interesting,' or words to that effect. But I had not the imagination to say, 'By golly, if this is true it opens up a whole new branch of physics.' And I think other physicists, with very few exceptions, at that time were as unimaginative as I."

Several physicists were prodded into action by the suggestions of Lee and Yang. The first to take up the gauntlet was Madam Chien-Shiung Wu, a professor of physics at Columbia University and widely regarded as one of the world's leading physicists. She was already famous for her work on weak interactions and for the care and elegance with which her experiments were always designed. Like her friends Yang and Lee, she, too, had been born in China and had come to the United States to continue her career.

The experiment planned by Madam Wu involved the beta-decay of cobalt-60, a highly radioactive isotope of cobalt which continually emits electrons. In the Bohr model of the atom, a nucleus of cobalt-60 may be thought of as a tiny sphere that spins like a top on an axis labeled north and south at the ends to indicate the magnetic poles. The beta-particles (electrons) emitted in the weak interaction of beta-decay are shot out from both the north and the south ends of nuclei. Normally, the nuclei point in all directions, so the electrons are shot out in all directions. But when cobalt-60 is cooled to near absolute zero (– 273 degrees on the centigrade scale), to reduce all the joggling of its molecules caused by heat, it is possible to apply a powerful electromagnetic field that will induce more than half of the nuclei to line up with their north ends pointing in the same direction. The nuclei go right on shooting out electrons. Instead of being scattered in all directions, however, the electrons are now concentrated in two directions: the direction toward which the north ends of the magnetic axes are pointing, and the direction toward which the south ends are pointing. If the law of parity is not violated, there will be just as many electrons going one way as the other.

To cool the cobalt to near absolute zero, Madam Wu needed the facilities of the National Bureau of Standards in Washington, D.C. It was there that she and her colleagues began their historic experiment. If the number of electrons divided evenly into two sets, those that shot north and those that shot south, parity would be preserved. The theta-tau puzzle would remain puzzling. If the beta-decay process showed a handedness, a larger number of electrons emitted in one direction than the other, parity would be dead. A revolutionary new era in quantum theory would be under way.

At Zurich, one of the world's greatest theoretical physicists, Wolfgang Pauli, eagerly awaited results of the test. In a now-famous letter to one of his former pupils, Victor Frederick Weisskopf (then at the Massachusetts Institute of Technology), Pauli wrote, "I do *not* believe that the Lord is a weak left-hander, and I am ready to bet a very high sum that the experiments will give symmetric results."

Whether Pauli (who died in 1958) actually made (like Feynman) such a bet is not known. If he did, he also lost. The electrons in Madam Wu's experiment were *not* emitted equally in both directions. Most of them were flung out from the south end; that is, the end toward which a majority of the cobalt-60 nuclei pointed their south poles.

At the risk of being repetitious, and possibly boring readers who see at once the full implication of this result, let us pause to make sure we understand exactly why Madam Wu's experiment is so revolutionary. It is true that the *picture* (Figure 62) of the cobalt-60 nucleus, spinning in a certain direction around an axis labeled *N* and *S*, is an asym-

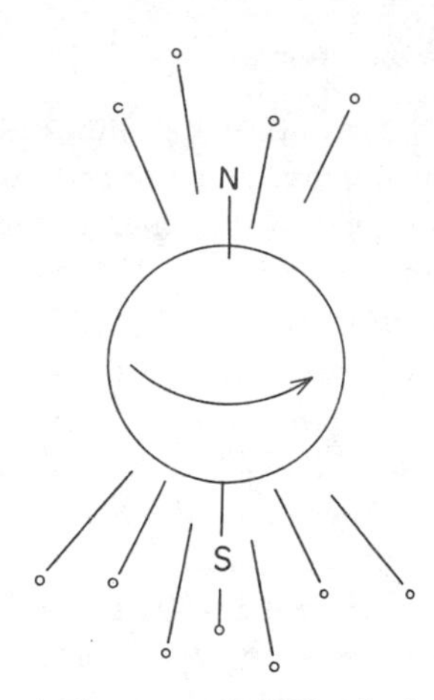

Figure 62. An electron is more likely to be flung out from the south end of a cobalt-60 nucleus than from its north end.

metric structure not superposable on its mirror image. But this is just a picture. As we have learned, the labeling of *N* and *S* is purely conventional. There is nothing to prevent one from switching *N* and *S* on all the magnetic fields in the universe. The north ends of cobalt-60 nuclei would become south, the south ends north, and a similar exchange of poles would occur in the electromagnetic field used for lining up the nuclei. Everything prior to Madam Wu's experiment suggested that such a switch of poles would not make a measurable change in the experimental situation. If there were some intrinsic, observable difference between poles—one red and one green, or one strong and one weak—then the labeling of *N* and *S* would be more than a convention. The cobalt-60 nuclei would possess true spatial asymmetry. But physicists knew of no way to distinguish between the poles except by testing their reaction to other magnetic axes. In fact, as we have learned, the poles do not really exist. They are just names for the opposite sides of a spin.

Madam Wu's experiment provided for the first time in the history of science a method of labeling the ends of a magnetic axis in a way that is not at all conventional. *The south end is the end of a cobalt-60 nucleus that is most likely to fling out an electron.*

The nucleus can no longer be thought of as analogous to a spinning sphere or cylinder. It must now be thought of as analogous to a spinning cone. Of course, this is no more than a metaphor. No one has the slightest notion at the moment of why or how one end of the axis is different, in any intrinsic way, from the other. But there is a difference! "We are no longer trying to handle screws in the dark with heavy gloves," was the way Sheldon Penman of the University of Chicago put it (*Scientific American*, July 1961); "we are being handed the screws neatly aligned on a tray, with a little searchlight on each that indicates the direction of its head."

It should be obvious now that here at long last is a solution to the Ozma problem—an experimental method of extracting from nature an unambiguous definition of left and right. We say to the scientists of Planet X: "Cool the atoms of cobalt-60 to near absolute zero. Line up their nuclear axes with a powerful magnetic field. Count the number of electrons flung out by the two ends of the axes. The end that flings out the most electrons is the end that we call 'south.' It is now possible to label the ends of the magnetic axis of the field used for lining up the nuclei, and this in turn can be used for labeling the ends of a magnetic needle. Put such a needle above a wire in which the current moves away from you. The north pole of this needle will point in the direction we call 'left.' "

We have communicated precisely and unambiguously to Planet X our meaning of the word *left*. Neither we nor they will be observing in common any single, particular asymmetric structure. We will be observing in common a universal law of nature. In the weak interactions, nature herself, by her own intrinsic handedness, has provided an operational definition of left and right. It is easy to understand why Pauli and other physicists did not expect Madam Wu's experiment to overthrow parity. It would have meant that nature is not ambidextrous.

In the context of my *Esquire* tale about left and right, the cobalt-60 experiment provides a method by which the puzzled astronauts could tell whether they were reversed. Of course, they would have to find some cobalt on the unknown planet, convert it to its radioactive isotope by bombarding it with neutrons, and so on. But assuming that they had the equipment and could find the necessary materials, they would be able to test their handedness.

Similarly, Madam Wu's experiment clearly violates the assertion that all natural events can be photographed on motion picture film and projected in reversed form without the viewer being the wiser.

EXERCISE 16: *Explain precisely how an observation of all details of the cobalt-60 experiment, when viewed as a projected motion picture, would enable one to tell whether the film had been reversed.*

Although evidence against the conservation of parity was strongly indicated by Madam Wu's work in late 1956, the experiment was not finally completed until early in January 1957. Results were formally announced by Columbia University's distinguished physicist Isador Rabi on January 15, 1957. The announcement also included the results of a confirming experiment conducted by Columbia physicists at the Nevis Cyclotron Laboratories at Irvington-on-Hudson in Westchester County, New York. This confirming test, made with mu mesons, showed an even stronger handedness. The mu mesons shot out twice as many electrons in one direction as in the other. Independent of both experiments, a third test was made at the University of Chicago using the decay of pi and mu mesons. It, too, showed violation of parity. All over the world, physicists began testing parity in other weak interactions. By 1958 it was apparent that parity is violated in *all* such interactions. The theta-tau puzzle was solved. There is only *one* K-meson. Parity is *not* conserved.

"A rather complete theoretical structure has been shattered at the base," declared Rabi (quoted by the *New York Times*, January 16, 1957), "and we are not sure how the pieces will be put together." An unnamed physicist was reported by the *Times* as saying that nuclear physics had been battering for years at a closed door only to discover suddenly that it wasn't a door at all—just a picture of a door painted on a wall. Now, he continued, we are free to look around for the true door. O. R. Frisch, the physicist who was a codiscoverer of nuclear fission, reports in his book *Atomic Physics Today* (Basic, 1961) that on January 16, 1957, he received the following air letter from a friend:

> Dear Robert:
> HOT NEWS. Parity is not conserved. Here in Princeton they talk about nothing else; they say it is the most important result since the Michelson experiment . . .

The Michelson experiment was the famous Michelson-Morley test in 1887, which established the constant velocity of light regardless of the motion of source and observer—a historic experiment that paved the way for Einstein's theory of relativity. Madam Wu's experiment may well prove to be equally historic.

The two tests were very much alike in their shattering element of surprise. Everybody expected Albert Michelson and Edward Morley to detect a motion of the earth relative to a fixed "ether." It was the negative result of this test that was so upsetting. Everybody expected Madam Wu to find a left–right symmetry in the process of beta-decay. Nature sprang another surprise. It was surprising enough that certain particles had a handedness; it was more surprising that handedness seemed to be observable only in weak interactions. Physicists felt a shock even greater than Mach had felt when he first encountered the needle-and-wire asymmetry.

"Now after the first shock is over," Pauli wrote to Weisskopf on January 27, after the staggering news had reached him, "I begin to collect myself. Yes, it was very dramatic. On Monday, the twenty-first, at 8 P.M. I was supposed to give a lecture on the neutrino theory. At 5 P.M. I received three experimental papers [reports on the first three tests of parity]. . . . I am shocked not so much by the fact that the Lord prefers the left hand as by the fact that he still appears to be left-handed symmetric when he expresses himself strongly. In short, the actual problem now seems to be the question: Why are strong interactions right-and-left symmetric?" [They are not completely. In 1967 the Russian physicist V. M. Lobashov found extremely minute violations of parity in the strong nuclear interactions.]

The Pakistani physicist Abdus Salam (from whose article on "Elementary Particles" in *Endeavor*, April 1958, the extracts from Pauli's letters are taken) tried to explain to a liberal-arts-trained friend why the physicists were so excited about the fall of parity. "I asked him," wrote Salam in this article, "if any classical writer had ever considered giants with only the left eye. He confessed that one-eyed giants have been described, and he supplied me with a full list of them; but they always sport their solitary eye in the middle of the forehead. In my view, what we have found is that space is a weak left-eyed giant."

Physicist Jeremy Bernstein, in an article on "A Question of Parity" that appeared in the *New Yorker*, May 12, 1962 (later reprinted in his book *A Comprehensible World*, Random House, 1967), reveals an ironic sidelight on the story of parity's downfall. In 1928 three physicists at New York University had actually discovered a parity violation in the decay of a radioactive isotope of radium. The experiment had been repeated with refined techniques in 1930. "Not only in every run," the experimenter reported, "but even in all readings in every run, with few exceptions," the effect was observable. But this was at a time when, as Bernstein puts it, there was no theoretical context in which to place these results. They were quickly forgotten. "They were," writes Bernstein, "a kind of statement made in a void. It took almost thirty years of intensive research in all branches of experimental and theoretical physics, and, above all, it took the work of Lee and Yang, to enable physicists to appreciate exactly what those early experiments implied."

In 1957 Lee and Yang received the Nobel Prize in physics for their work. Lee was then thirty, Yang thirty-four. The choice was inevitable. The year 1957 had been the most stirring in modern particle physics, and Lee and Yang had done most of the stirring. If you are curious to know more about these two remarkable men, look up Bernstein's excellent article.

It is worth pausing to note that, like so many other revolutions in physics, this one came about as the result of largely abstract, theoretical, mathematical work. Not one of the three experiments that first toppled parity would have been performed at the time it was performed if Lee and Yang had not told the experimenters what to do. Lee had had no experience whatever in a laboratory. Yang had worked briefly in a lab at the University of Chicago, where he was once a kind of assistant to the great Italian physicist Enrico Fermi. He had not been happy in experimental work. His associates had even made up a short rhyme about him, which Bernstein repeats:

> Where there's a bang,
> There's Yang.

Laboratory bangs can range all the way from an exploding test tube to the explosion of a hydrogen bomb. But the really big bangs are the bangs that occur inside the heads of theoretical physicists when they try to put together the pieces handed to them by the experimental physicists.

John Campbell, Jr., the editor of *Analog Science Fiction*, once speculated in an editorial that perhaps there was some difference in the intellectual heritage of the Western and Oriental worlds which had predisposed two Chinese physicists to question the symmetry of natural law. It is an interesting thought. I myself pointed out, in my Mathematical Games column in *Scientific American*, March 1958, that the great religious symbol of the Orient (it appears on the Korean national flag) is the circle divided asymmetrically as shown in Figure 63. The dark and light areas are known respectively as the Yin and Yang. The Yin and Yang are symbols of all the fundamental dualities of life: good and evil, beauty and ugliness, truth and falsehood, male and female, night and day, sun and moon, heaven and earth, pleasure and pain, odd and even, left and right, positive and negative—the list is endless. This dualism was first symbolized in China by the odd and even digits that alternate around the perimeter of the *Lo shu*, the ancient Chinese magic square of order 3. Sometime in the tenth century the *Lo shu* was replaced by the divided circle, which soon became the dominant Yin-Yang symbol. When it was printed or drawn, black and white was used, but when painted, the Yang was made red instead of white. The two small spots were (and still are) usually added to symbolize the fact that on each side of any duality there is always a bit of the other side. Every good act contains an element of evil, every evil act an element of good; every ugliness includes some beauty, every beauty includes some ugliness, and so on.[2] The spots remind the scientist that every "true" theory contains an element of falsehood. "Nothing is perfect," says the Philosopher in James Stephens's *The Crock of Gold.* "There are lumps in it."

EXERCISE 17: *There is a three-dimensional analog of the Yin-Yang, so familiar that almost everyone has at one time held a model of it in his hands. What is it? Is it left–right symmetrical?*

Figure 63. The asymmetric Yin-Yang symbol of the Orient.

The history of science can be described as a continual, perhaps never-ending, discovery of new lumps. It was once thought that planets moved in perfect circles. Even Galileo, although he placed the sun and not the earth at the center of the solar system, could not accept Kepler's view that the planetary orbits were ellipses. Eventually it became clear that Kepler had been right: the orbits are *almost* circles but not quite. Newton's theory of gravity explained why the orbits were perfect ellipses. Then slight deviations in the Newtonian orbits turned up and were in turn explained by the correction factors of relativity theory that Einstein introduced into the Newtonian equations. "The real trouble with this world of ours," comments Gilbert Chesterton in *Orthodoxy*, "is not that it is an unreasonable world, nor even that it is a reasonable one. The commonest kind of trouble is that it is nearly reasonable, but not quite. . . . It looks just a little more mathematical and regular than it is; its exactitude is obvious, but its inexactitude is hidden; its wildness lies in wait."

To illustrate, Chesterton imagines an extraterrestrial examining a human body for the first time. He notes that the right side exactly duplicates the left: two arms, two legs, two ears, two eyes, two nostrils, even two lobes of the brain. Probing deeper, he finds a heart on the left side. He deduces that there is another heart on the right. Here, of course, he encounters a spot of Yin within the Yang. "It is this silent swerving from accuracy by an inch," Chesterton continues, "that is the uncanny element in everything. It seems a sort of secret treason in the universe. . . . Everywhere in things there is this element of the quiet and incalculable."

Feynman, with no less reverence than Chesterton, says the same thing this way at the close of a lecture on symmetry in physical laws (lecture 52 in *The Feynman Lectures on Physics*, Addison-Wesley, 1963):

"Why is nature so nearly symmetrical? No one has any idea why. The only thing we might suggest is something like this: There is a gate in Japan, a gate in Neiko, which is sometimes called by the Japanese the most beautiful gate in all Japan; it was built in a time when there was great influence from Chinese art. This gate is very elaborate, with lots of gables and beautiful carvings and lots of columns and dragon heads and princes carved into the pillars, and so on. But when one looks closely he sees that in the elaborate and complex design along one of the pillars, one of the small design elements is carved upside down; otherwise the thing is completely symmetrical. If one asks why this is, the story is that it was carved upside down so that the gods will not be jealous of the perfection of man. So they purposely put the error in there, so that the gods would not be jealous and get angry with human beings.

"We might like to turn the idea around and think that the true explanation of the near symmetry of nature is this: that God made the laws only nearly symmetrical so that we should not be jealous of His perfection!"

Note that the Yin-Yang symbol is asymmetrical. It is not superposable on its mirror image. The Yin and Yang are congruent shapes, each asymmetrical, each with the same handedness. By contrast, the Christian symbol, the cross, is left–right symmetrical. So is the Jewish six-pointed Star of David, unless it is shown as an interlocking pair of triangles that cross alternately over and under each other. It is a pleasant thought that perhaps the familiar asymmetry of the Oriental symbol, so much a part of Chinese culture, may have played a subtle, unconscious role in making it a bit easier for Lee and Yang to go against the grain of scientific orthodoxy; to propose a test which their more symmetric-minded Western colleagues had thought scarcely worth the effort.

Notes

1. For the benefit of readers interested in recreational mathematics, I cannot resist adding that Feynman is one of the codiscoverers of hexaflexagons, those remarkable paper-folded structures that keep changing their faces when flexed. (See chapter 1 of my *Scientific American Book of Mathematical Puzzles and Diversions.*) Although a hexaflexagon *looks* perfectly symmetrical, its inner structure possesses a handedness; that is, any given flexagon can be constructed in either a left- or right-handed way.

2. For these facts about the Yin-Yang symbol I am indebted to Schuyler Cammann's excellent article on "The Magic Square of Three

in Old Chinese Philosophy and Religion," *History of Religions*, vol. 1, no. 1, summer 1961, pp. 37–80.

23. NEUTRINOS

The famous Michelson-Morley experiment was performed in 1887. It was not until 1905, some eighteen years later, that the full implications of the experiment came to light in Einstein's first paper on the special theory of relativity. No one knows how many years will go by before the full implications of Madam Wu's experiment are spelled out by another Einstein.

At the moment, the world's leading mathematical physicists are doing the best they can to develop a general theory that could account for the violation of parity. Hardly a month passes that papers containing such explanations are not received by the physics journals. Most of the papers, alas, are written by physicists and engineers who have rushed into theory without bothering to learn all the facts, and especially without troubling to learn the difficult mathematics of quantum theory. Nevertheless, no one should rule out the possibility that some amateur, lacking the expert knowledge that might make him overcautious, may stumble on a new insight or gimmick that will unlock a genuine door.

One startling thought occurred immediately to everybody in the field: could space itself possess at every point some sort of intrinsic handedness? Both the classical physics of Newton and the equations of modern relativity theory and quantum theory assume that space is completely isotropic. This means that one direction in space is no different from another; space is spherically symmetrical. Is it possible to construct models of the cosmos in which space has an intrinsic handedness?

Yes, mathematicians can construct models of anisotropic (not isotropic) 3-space that have an asymmetry of the same handedness at every point. It isn't easy, and such spaces are far from simple. You might think that an overall twist of space, comparable to the twist of a

Moebius strip, would do the trick, but it doesn't. The twist has to be present at every point and be of such a character that its effect on weak interactions is the same regardless of the orientation of the apparatus. Since the earth spins in space, the apparatus used in parity tests is constantly changing its orientation, yet the test results remain constant. One has to construct a space in which there is some sort of fine, unobservable "grain" that provides a uniform asymmetric twist regardless of the orientations of the particles affected by the twist.

Assuming that such a "grain" exists, it is not hard to see how parity might be strongly violated only in weak interactions. In stronger forces the subtle, minute twist of space would be negligible. If you are bowling on a warped lane, the effect of the warp can be overcome by bowling a fast ball. In fact, you can give the ball a spin that will make it curve *against* the warp. But if you release a slow-moving ball, or use a ball as small as a pea, the warp in the lane is more likely to distort its path. In a similar way the strong, fast-moving interactions of particles may tend to eliminate the effect of an asymmetric grain in space-time. Large macroscopic movements of billiard balls and planets, and radiation moving with the speed of light, might similarly overcome the effect of such a grain.

This approach to the problem has appealed to a number of physicists. It is favored, for example, by Otto Frisch, of Cambridge, in his book (cited earlier) *Atomic Physics Today*. "May we suppose," he asks, "that cobalt would not be radioactive if space were not twisted?" In spite of the attractiveness (in some ways) of this theory, most particle physicists would, I think, answer no.

For one thing, gravity is much weaker than the force involved in weak interactions, and known to be intimately involved with the space-time structure of the cosmos. One would expect asymmetry to show up in gravitational effects, but such asymmetry has never been detected. It is true that gravity is so weak that it is completely negligible on the particle level, but if the theory of general relativity is true, gravity is only another way of talking about inertia. Particles do possess inertial mass. In all experiments so far made with elementary particles, nothing has been observed that suggests the slightest trace of inertial asymmetry, a fact extremely difficult to reconcile with twisted points in space-time. For these reasons physicists are understandably reluctant to give up the classic notion that space is isotropic.

Fortunately, there is a second approach to the weak interactions in which handedness can be explained without having recourse to a twist in the grain of space-time. This approach rests on the assumption that somehow—exactly how remains a total mystery—some of the elementary particles exhibit in their structure a true spatial asymmetry. We have seen how chemists once found it hard to believe that molecules could possess an asymmetric structure in the way their atoms are linked, but the discovery of stereoisomers finally cleared up the mystery. Many physicists think that our present knowledge of elementary particles is on a par with the knowledge of molecules before the discovery of stereoisomerism. It has not been necessary, Teller reminds us, to investigate the internal structure of the electron "yet." Is it possible that in the future, with the aid of tools now unimaginable, physicists will find that the elementary particles are far from elementary?

At present there are only vague, tantalizing hints in this direction. The strongest hint has come from the recent discovery that the neutrino is, indeed, a structure with a true spatial handedness.

The neutrino's history is worth sketching. As mentioned earlier, the neutron (present in the nuclei of all atoms except hydrogen) is a particle with a magnetic moment but zero charge. It has a mass slightly more than the proton's mass. In the beta-decay of radioactive nuclei, a neutron breaks down into a proton and an electron. However, the mass of the proton and electron, added together, do not quite equal the mass of the original neutron. Some of the missing mass has been converted to energy in accordance with Einstein's formula $E = mc^2$. Even when this is taken into account, however, there is still a slight amount of mass-energy unaccounted for. Where does it go? In 1931 Pauli suggested that it is carried away by an invisible, undetected "thief"—an elusive particle whose existence had to be assumed in order to balance the two sides of the equation. When Fermi developed his theory of weak interactions to explain the slowness of beta-decay, he took over Pauli's suggestion. The "neutrino," or "little neutral one," was Fermi's happy choice of a name for Pauli's thief particle. The properties of the neutrino were such that it seemed impossible to trap it. But its reality was finally established in 1956 by Frederick Reines and Clyde L. Cowan, Jr., using as their source of neutrinos the Atomic Energy Commission's huge nuclear pile on the

Savannah River, in Georgia. It later turned out that what they caught were antineutrinos, which are produced in great abundance by nuclear fusion reactors, but this is getting ahead of our story.

Years ago, in an animated color cartoon, there was a song with a refrain that went: "You're nothin' but a nothin', you're not a thing at all." The neutrino is about as close to this description as an elementary particle can be and still be something. Its rest-mass is believed to be zero. If so, it must travel through space with the speed of light. It has no charge, no magnetic field. It does have "spin." In fact, that is about all it has. It is, as some physicists have put it, almost pure spin, like the pure grin of the Cheshire Cat.

Because it is neither attracted nor repelled by the electrical and magnetic fields of other particles, a neutrino coming from outer space is likely to go clean through the earth just as if nothing at all were in the way. The chance of its being stopped by an earth particle is estimated at about one in ten billion. Fortunately, there are enough neutrinos around so that collisions *do* occur, otherwise the little neutral one would never have been detected. As you read this sentence, billions of neutrinos, coming from the sun and other stars, perhaps even from other galaxies, are streaming through your skull and brain. John Updike, in his poem "Cosmic Gall," expressed it this way:

> Neutrinos, they are very small.
>   They have no charge and have no mass
> And do not interact at all.
> The earth is just a silly ball
>   To them, through which they simply pass,
> Like dustmaids down a drafty hall
>   Or photons through a sheet of glass.
>   They snub the most exquisite gas,
> Ignore the most substantial wall,
>   Cold-shoulder steel and sounding brass,
> Insult the stallion in his stall,
>   And, scorning barriers of class,
> Infiltrate you and me! Like tall
> And painless guillotines, they fall
>   Down through our heads into the grass.

At night, they enter at Nepal
And pierce the lover and his lass
From underneath the bed—you call
It wonderful; I call it crass.

The harmlessness of the neutrino prompted Ralph S. Cooper, a young physicist at the Los Alamos Scientific Laboratory, to invent in 1961 a wonderful new weapon which he called the neutrino bomb. You may recall that in 1961 there was considerable talk of developing a "clean" neutron bomb that would have no heat or blast effects and would leave no radioactive fallout. Buildings would be left intact. The bomb, with a great burst of neutron radiation, would do no more than destroy all life within its range. In 1977 the possibility of the United States making neutron bombs was revived, and the debate is still going on about whether this would be a wise or a foolish thing to do. Cooper's proposal was to make an even cleaner bomb that would produce a great burst of neutrino radiation. The penetrating power of neutrinos is so much greater than that of neutrons, Cooper pointed out, that they make the neutrons look like marshmallows. A neutrino bomb would, he said, be the "ultimate in clean, blastless, nuclear weapons."

Cooper's spoof first appeared in the *Los Alamos Scientific Laboratory News*, July 13, 1961, and was reprinted in Groff Conklin's *Great Science Fiction by Scientists* (Collier Books, 1962). It is too good to pass by without giving a few more details. The bomb's charge would consist of hydrogen, but hydrogen with its protons and electrons converted by an ingenious procedure into two new particles, the pseudo-proton and pseudo-electron. "The pseudo-electron would have no spin," reported the *New York Times*, August 13, 1961, "and no strangeness (a property of elementary particles). It would be called a 'fiction.' The pseudo-proton would also have no spin but it would possess a strangeness of one. It would be called a 'truth' particle or 'truthiton,' truth being stranger than fiction. The two pseudo-particles would annihilate each other in a complicated interaction in which a new element, called truthitonium, would be formed. Each atom of truthitonium would decay radioactively into 2,000 neutrinos in a time that Dr. Cooper would call a 'moment of truth.' "

"Once the neutrino bomb is detonated," explained Cooper, "there is not one particle of truth left in it." The detonation (caused by air rushing into the temporary vacuum produced by the disappearance of the hydrogen) would produce, said Cooper, "a loud bang, informing the victims in the target area that they have been had."

Assuming that the neutrino has spin and moves in a direction parallel with the spin axis, it is obvious that it can spin in either of two ways. Imagine a point on the outside of such a particle. (Of course, this is the crudest possible way to talk about something that can be described accurately only by mathematical formulas, but the crude analogy is not entirely meaningless.) As the particle moves forward with the speed of light, the point will generate either a right or a left helix. When we talk of it as right or left, this assumes that the observer is at rest or moving at a slower speed than the particle, relative to some outside frame of reference. If the observer is moving faster than the particle, and in the same direction, then the particle has a relative motion that is *away* from the observer. This would reverse the handedness of the particle's helix.

To understand this, imagine that a neutrino with a right-handed helical motion is approaching you. You will see it from its front end, so to speak, as a right-handed helix. It passes you and moves away. You see it from its back end, but it is still a right-handed helix. Suppose now that you and the right-handed neutrino are traveling in the same direction, but you are going twice as fast as the neutrino. There are no absolute motions in space-time, only relative ones. From your frame of reference, which is just as good and true as any other frame of reference (there are no "preferred" frames of reference in relativity theory), the neutrino will be moving *away* from you. You will see it as a *left*-handed helix. The same would be true if you were behind the neutrino and gaining on it. Relative to an outside frame of reference such as the fixed stars, you would be gaining on a right-handed neutrino, but relative to *your* frame of reference the neutrino would be a *left*-handed helix moving toward you.

Can a neutrino, then, be either left- or right-handed, depending on its motion relative to the observer? No, because the neutrino, like the photon of light radiation, moves with the speed of light. Relativity theory does not permit an observer to move that fast. For this reason he must always see a particular neutrino, whether moving toward him or away from him, as having the same handedness. He cannot attach himself to a frame of reference from which he sees the neutrino differently. In short, the handedness of a neutrino is invariant for all possible observers.

The suggestion that a spinning particle can be stable in either of two helical forms, one a mirror image of the other, was advanced by Hermann Weyl, the great German mathematician, as early as 1929. He had absolutely no experimental data on which to base this speculation; it just seemed to him that simplicity and mathematical beauty demanded it. No one paid much attention to Weyl's theory. Why? Because it violated the conservation of parity. It introduced into nature an inexplicable asymmetry. The instant parity was overthrown, Weyl's theory became a prophetic guess. Indeed, evidence quickly accumulated that the neutrino had an antiparticle, the antineutrino, and that the two could be distinguished in just the way Weyl had suggested.

This two-component theory of the neutrino, as it is called, was advanced independently in 1957 by a number of theoretical physicists: Lee and Yang, Abdus Salam of Pakistan, and Lev Davidovich Landau of the U.S.S.R. (Weyl died in 1955, two years before his theory was revived.) There is strong evidence that the theory is essentially correct. In beta-decay, when electrons are flung out from nuclei, they are accompained by antineutrinos that have a clockwise spin as seen from the nucleus—that is, their path has the twist of a right-handed helix. On the other hand, when an antineutron breaks down in the process of antibeta-decay, positrons are flung out, accompanied by left-handed neutrinos (see Figure 64). Here for the first time in the history of particle physics a particle has been shown to have a stable

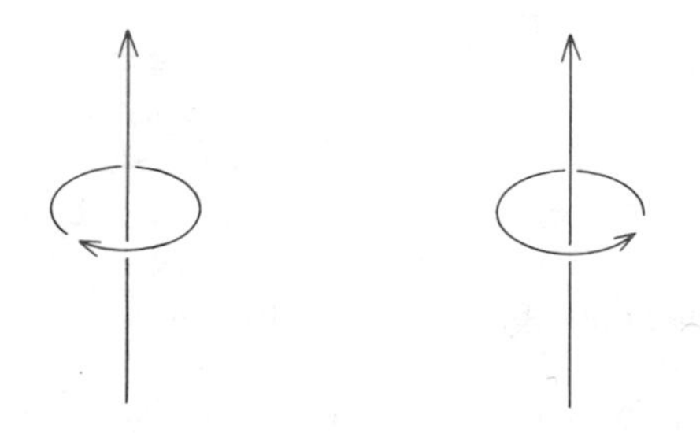

Figure 64. Diagrams of the neutrino (left) and antineutrino (right).

asymmetric structure (in this case the structure is simply a union of the spin and direction of travel) that exhibits true spatial asymmetry. The neutrino and antineutrino are the first known analogs, on the particle level, of Pasteur's left- and right-handed tartaric acid molecules.

In 1957 several physicists, including Lee and Yang, complicated the picture. They suggested there might be two different kinds of neutrino-antineutrino pairs, one associated with decays in which electrons are emitted, the other with decays in which muons are emitted. This was confirmed in 1962 by a team of physicists from Columbia University and the Brookhaven National Laboratory, working with the Alternating Gradient Synchrotron, at Brookhaven, in Yaphank, Long Island, New York.

The muon, discovered in 1936, is one of the most tantalizing mysteries in particle physics. It is identical with the electron in all its interactions but has a mass more than 200 times that of the electron. It is as if the electron, under certain conditions, can become 200 times heavier than usual.

Is the muon a fat electron or a different particle? Physicists are mystified because they can't think of any reason why it should exist. Their attitude is like that of the Wasp in the recently found "lost" episode of *Through the Looking-Glass.* The Wasp considered Alice's eyes so close together that he wondered why she had two when one eye would do just as well.

The mystery is not lessened by the fact that the muon has its own neutrino, indistinguishable from the electron neutrino except that when it is produced by an interaction involving muons it "remembers" that it belongs to a muon, so when it reacts with a proton or neutron it produces another muon. The electron muon similarly remembers what it is.

Exactly what is going on here is far from clear. Do the two kinds of neutrinos spin the same way, with their antiparticles spinning the opposite way, or does each neutrino spin the same way as the other neutrino's antiparticle? This question has yet to be decided. An unnamed physicist was quoted by the *New York Times,* July 1, 1962, as exclaiming with wonderment: "It is as though we had discovered two kinds of vacuum!"

Neutrinos are the only particles produced by thermonuclear reactions inside of stars that can easily escape. Billions upon billions of these extraterrestrial electron neutrinos are constantly streaming through the earth, all but a minute portion coming from our sun. It is estimated that 3 percent of the energy radiated by the sun consists of electron neutrinos.

For many years Raymond Davis, Jr., has been trapping solar neutrinos with a detector tank that contains 100,000 gallons of a common cleaner fluid. It is about a mile deep inside a gold mine in Lead, South Dakota. The results have been puzzling. After six years of work he got at the most only about one-fifth of the anticipated number of neutrinos, although later the proportion went up considerably.

Physicists are still arguing about what all this means. Does the sun fluctuate in the number of neutrinos it generates? Will standard theories about solar interactions have to be revised? Is it possible that neutrinos decay in transit? The last possibility seems remote because it would mean that the neutrino has a slight mass, and there is no good reason yet to suppose it has. The case of the missing neutrinos is still a dark mystery waiting to be unraveled.

24. MR. SPLIT

When an electron meets a positron, the masses of the two particles vanish in a burst of radiation. We have seen (in chapter 21) how Dirac once explained this, as well as simultaneous pair creation, in terms of a "hole" theory. A particle taken from a dense continuum leaves behind a hole that is its antiparticle. When a particle drops back into the hole, particle and hole vanish. More recently, in a fascinating article on "The Physicist's Picture of Nature" in *Scientific American* (May 1963), Dirac suggested a different picture. He likened the electron and positron to the two *ends* of an electromagnetic line of force. The line has a direction that serves to distinguish its two ends. The meeting of electron and positron would be comparable to joining the plus end of one string to the minus end of another. The ends (electron and positron) vanish, leaving only a line of force. Similarly, cutting a line of force would result in pair creation of plus and minus ends.

Such pictures cannot, of course, be taken literally; they are only suggestive of theories that have to be worked out mathematically and tested experimentally. They are attempts to account for what at the moment is the greatest of all mysteries in quantum theory: the nature of the electric charge. No one knows what distinguishes a positive from a negative charge, why charge is always an exact multiple of one quantum, or why a quantum of negative charge has exactly the same strength as a quantum of positive charge. These mysteries are obviously bound up with the simultaneous pair creation and pair annihilation of a particle and its antiparticle.

Is there a picture that will explain positive and negative charge in terms of left- and right-handedness? Yes, it is easy to think of many crude pictures along such lines. A few years ago I read in a newspaper that a London official had once again proposed that a bridge be built across the English Channel to link Britain and France. Since British cars travel on the left side of the road, French cars on the right, I instantly had a vision comparable to a particle–antiparticle clash. Surely traffic on the bridge would come to a dead standstill. A better picture is furnished by the meeting of two smoke rings with opposite vortex motions. The spin momenta naturally cancel each other and the two structures dissolve.

In plane geometry a simple picture of pair creation and annihilation is provided by the equilateral triangle. It is, of course, symmetrical—identical with its mirror image. Bisect it vertically from the base (Figure 65) and you witness pair creation of two asymmetric right triangles of opposite handedness. Neither triangle is superposable on the other without taking it out of 2-space and turning it over. Bring two triangles of opposite handedness together, reforming an equilateral triangle, and there is simultaneous pair annihilation.

Imagine the plane of 2-space covered with tiny triangles, some equilateral, some the left sides of equilateral triangles, some the right sides. It is a picture curiously similar to that of positive and negative charges in the universe. The law of "conservation of charge," which has never been found violated, asserts that the net amount of charge in the cosmos never alters. This is true also of our triangles. Suppose we start with 1,000 "neutral" equilateral triangles, 500 "negative" (right) sides of such triangles, and 200 "positive" (left) sides. There are 300

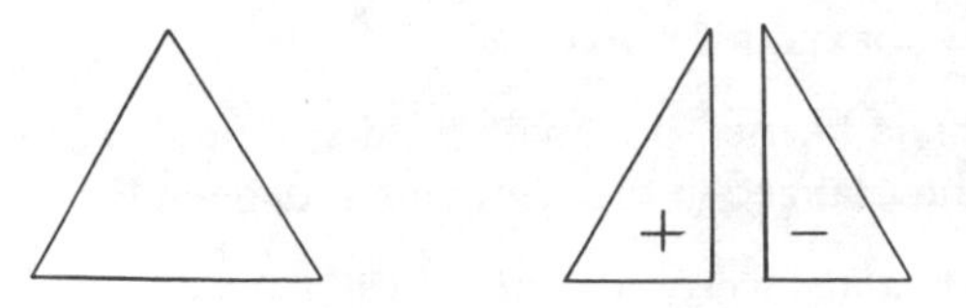

Figure 65. A simple model of charge conservation.

more right sides than left. This is the net charge of the system. We can split as many equilateral triangles in half as we please, and put any number of halves together again, but because we must create and destroy in pairs there always will be precisely 300 more right-handed triangles than left. The net charge of the system is conserved.

A whimsical 3-space picture of the same situation is provided by L. Frank Baum in one of his little-known non-Oz fantasies, *Dot and Tot of Merryland* (George M. Hill, 1901). The sixth valley of Merryland is inhabited by wind-up animals, cars, and other toys. The toys are kept wound by an overseer called Mr. Split. Mr. Split has such a heavy work load that, when the going gets tough, he splits himself exactly in half down the middle. Each half hops about separately, on its single leg, winding up toys. Mr. Left Split, who is bright red, speaks only the left halves of words. Mr. Right Split, who is white, speaks only the right halves of words. When the two hook themselves together, Mr. Split speaks normally. "I do not think there is another man in the whole world," says the queen of Merryland, "that does so much work as Mr. Split."

Here again, if we think of positive and negative charges as the two halves of endless numbers of Mr. Splits, sometimes working separately, sometimes hooked together, we have another analogy with the law of charge conservation.[1]

Conjurors know many tricks with rope, string, and handkerchiefs (twisted in rope fashion) that provide entertaining examples of mutual annihilation by a left–right encounter. In most cases the two structures involved are helices of opposite handedness. Charles Howard Hinton, a somewhat eccentric American mathematician who married one of the daughters of the English mathematician George Boole, described one such trick to illustrate his theory of positive and negative charge. In the first chapter of his book *A Picture of Our Universe* (it is reprinted in the first series of *Scientific Romances,* Allen &

Unwin, 1888) he likened positive and negative charge to the fabled Irish cats, immortalized in the anonymous doggerel:

> There once were two cats of Kilkenny,
> Who thought there was one cat too many,
>   So they mewed and they bit
>   And they scratched and they fit,
> Till, excepting their nails and the tips of their tails,
> Instead of two cats there weren't any.

"It is perfectly possible," wrote Hinton, "to make a model of the Kilkenny cats. And I propose to symbolize . . . the Kilkenny cat by a twist."

Hinton's model was a piece of rope twisted around a stick in the manner shown in Figure 66. First wind the rope several times around the stick, then hold the rope in place with your left thumb so you can continue winding, but in the opposite direction. Stop after you have made the same number of turns that you made the other way. If you imagine a plane of symmetry bisecting the stick, it is obvious that the coil on one side is the mirror image of the coil on the other. Release your thumb's pressure on the center of the rope and pull on the two ends that hang down. The rope will pull free of the stick. Each helix is a Kilkenny cat. The pull on the rope is their battle. Because the twists are mirror images, they annihilate each other.

"This is the mechanical conception I wish you to adopt," Hinton said. "There are such things as twists. Suppose by some means to every twist there is produced its image twist. These two, the twist and its image, may exist separately; but suppose that whenever a twist is produced its image twist is also produced, and that these two when

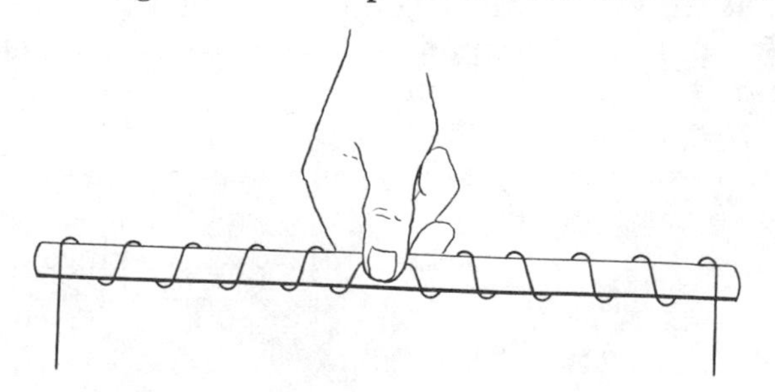

Figure 66. Charles Howard Hinton's rope trick.

put together annihilate each other. With this conception let us explore the domain of those actions which are called electrical."

Hinton proceeded to apply this picture to the meeting of two opposite electric charges. Each charge is a wave motion of helical character, but the twist is not in 3-space; it is in a higher space. A positively charged particle is, in some way unknown and inconceivable to us, a mirror image of a negatively charged particle. It possesses momentum in the higher space. For this reason, in accordance with the law of action and reaction, the production of a charged particle demands the simultaneous creation of its mirror twin. "No body can be made to move in any direction without imparting an equal motion in an opposite direction to another body," he wrote, "*e.g.*, the motion of a cannon ball is equalled by the recoil of the cannon. And so no twist can be given to the particles of a body without an image twist being given to other particles."

If we think of macroscopic bodies as made of particles possessing charge, then the mirror reversal of a body would result in what physicists call "charge conjugation," a change of all plus charges to minus and all minus to plus. Let me quote in full Hinton's most remarkable passage:

"If we consider a twist and its image, they are but the simplest and most rudimentary type of an organism. What holds good of a twist and its image twist would hold good of a more complicated arrangement also. If a bit of structure apparently very unlike a twist, and with manifold parts and differences in it—if such a structure were to meet its image structure, each of them would instantly unwind the other, and what was before a complex and compound whole, opposite to an image of itself, would at once be resolved into a string of formless particles. A flash, a blaze, and all would be over.

"To realize what this would mean we must conceive that in our world there were to be for each man somewhere a counterman, a presentment of himself, a real counterfeit, outwardly fashioned like himself, but with his right hand opposite his original's right hand. Exactly like the image of the man in a mirror.

"And then when the man and his counterfeit meet, a sudden whirl, a blaze, a little steam, and the two human beings, having mutually unwound each other, leave nothing but a residuum of formless particles."

Dr. Teller meets Dr. Anti-Teller!

We must remember that Hinton was writing about positive and negative charge, not about particles and antiparticles (the concept of antimatter did not come until many decades later). Nevertheless, his words have a curiously prophetic ring. When a neutrino and antineutrino meet in mutual suicide, surely something analogous to Hinton's rope trick takes place. Since neutrinos are involved in every weak interaction, is it possible that violations of parity may be due, in some unknown way, to the influence of the neutrino's asymmetry? Do neutrinos act as midwives in such effects, telling the electrons which way to go and which way to spin? Or is there a more fundamental asymmetry involved that is somehow bound up with electric charge?

Hinton's theory of positive and negative charge is not as crazy as it sounds. In 1921 the German physicist Theodore Kaluza developed what is known as the five-dimensional theory of relativity. Five years later Oscar Klein, then in Stockholm, extended Kaluza's theory in such a way as to suggest an explanation of positive and negative charge that has a striking resemblance to Hinton's picture.

It is impossible to explain Klein's theory adequately without mathematical equations, but in a rough way it amounts to this. In addition to the four space-time dimensions of orthodox relativity theory there is a fifth dimension, spatial in character. This fifth dimension curves back on itself like the surface of a cylinder, or, rather, like the surface of an incredibly slender thread, because the radius of curvature is very much smaller than the radius of an atom. (Klein estimated the radius as 10^{-30} cm., which means 1/1000000000000000000000000000000 centimeter.) Macroscopic objects are confined to four-dimensional space-time, but elementary particles have what physicists call a higher degree of freedom: we can think of particles as capable of moving around this fifth coordinate in either direction. If they go around one way, they are positively charged; if they go the other way, they are negatively charged.

Uncharged particles move along geodesics (straightest possible paths) through four-dimensional space-time. We can think of charged particles as tracing helical geodesics through five-dimensional space-time. Two oppositely charged particles, traveling a collision course, will have "world lines" of opposite handedness. When they meet, their respective momenta, being in opposite directions, cancel each other and both charges vanish. Similarly, if a neutral particle acquires a charge, its momentum exerts a recoil on another particle. The result is two helical world lines of opposite handedness as the two oppositely charged particles separate.

Let me add quickly that Klein's theory won few adherents, though it prompted a flurry of discussion in the late 1920s.[2] Einstein himself was interested in it for a time, but eventually decided against it. I mention it here only to show that Hinton's crude prerelativity picture *can* be given sophisticated mathematical formulation. There have been many field theories of relativity that have viewed positive and negative charges as enantiomorphic twins (for example, the theory first outlined by Sir Arthur Eddington in his *Relativity Theory of Protons and Electrons,* Cambridge University Press, 1936). So far, no such theory has been found satisfactory. Among some physicists, however, there is a growing suspicion that the mirror-image picture of charge may yet prove to be, like the little white spot of Yang inside the blackness of the Yin, a spot of truth inside the overall falseness of such theories.

The neutrino and its antiparticle are genuine mirror images of each other. Is it possible that somehow—perhaps in terms of a space and time wildly unlike the space and time we know on the macroscopic level—*every* particle is a true mirror image of its antiparticle? Is it possible that antimatter is nothing more than ordinary matter with its entire space-time structure, down to the last detail, reversed as by a looking glass?

No physicist is willing to give a firm yes to these questions; nevertheless, some evidence points in that direction. A number of experiments since 1957 have indicated that if the substances used in any parity-violating effect are charge-reversed (positive charges becoming negative and vice versa), then the handedness of the experiment also switches. We can make this clear by imagining a large vertical mirror on the wall of the room where Madam Wu made her experiment. The emission of the electrons in the experiment has a handedness that is reversed by this mirror. The experiment is not superposable on its mirror image. But if we imagine that on the other side of the mirror Madam Anti-Wu performs the same experiment, with equipment and substances made of antimatter, then the electrons will "go the other way." The symmetry of the basic laws of nature is restored.

Yang has put it this way. If we define mirror reflection as a left–right reversal *plus a charge reversal,* symmetry is preserved. This restoration of symmetry by double inversion (space and charge) has also been stressed by Eugene Wigner at Princeton and L. D. Landau in the U.S.S.R.[3] Of course, as Yang fully realizes, his is only a verbal statement that tells us nothing at all about *why* the addition of charge inversion to mirror inversion should restore symmetry.

If—and it is a gigantic if—the difference between positive and negative charge should turn out to rest on a simple left–right distinction of some sort, as Hinton suggested, then the new type of mirror reflection defined by Yang would turn out to be just a plain, old-fashioned, familiar mirror reversal. "It is easy to see," writes Landau (in the paper cited in the footnote), "that invariance of the interactions with respect to combined inversion leaves space completely symmetrical, and only the electrical charges will be asymmetrical. The effect of this asymmetry on the asymmetry of space is no greater than that due to chemical stereoisomerism." This would go far toward explaining the mystery of charge conservation. A unit of charge, if it were a stable asymmetric structure, could no more change to an opposite charge than a left-handed neutrino could turn into its right-handed twin. It could vanish only by combining with its twin. It could be created only in combination with its twin.

If antimatter is reflected matter, then we can answer Alice's question, "Is Looking-glass milk good to drink?" with a thunderous "No!" Such milk, touched by the hand or lips of an Alice, would create an explosion greater than that of the hydrogen bomb. H. G. Wells's unfortunate Mr. Plattner, who got turned over in 4-space, would have been annihilated the instant he tumbled back into this world. The astronauts in my *Esquire* story would have had no need of making a parity test to determine whether they had been reversed. If they had been, their spaceship would have exploded the instant they landed on the planet.

At this point the interested reader must steel himself against an almost irresistible urge to invent mirror-inversion theories of antimatter; theories based on the crude image of a spinning ball.

Nothing is easier than to devise such naive theories. You have only to imagine, say, that one end of an axis of rotation is itself revolving in a small circle and you immediately create a picture in which four types of spinning spheres can be distinguished: the spin of the axis's end can be with or against the spin of the sphere, and each of these two cases has its mirror image. Are these pictures of the four neutrinos? The answer is an unqualified no. The spinning-ball picture is no substitute here for quantum mechanics, and nothing but wasted time results from rushing in where informed mathematical physicists fear to tread.

How Pasteur would have exulted in the fall of parity! As we saw in chapter 16, he had a strong intuitive hunch that a fundamental handedness pervades the structure of the universe, and he spent many years trying to prove it. Today's biochemists no longer feel, as did Pasteur, that one need look this deep for an explanation of the asymmetry of organic molecules. There are simpler, more plausible explanations that do not invoke the asymmetry of elementary particles or a twist of space itself. Nevertheless, one cannot completely rule out the possibility that whatever is responsible for the asymmetry of weak interactions may also play a role in the formation of primitive organic compounds. Perhaps it will not be long before the exploration of other planets will throw light on this question. If, for example, astronauts find right-handed amino acids on Mars (instead of left-handed, as on earth), it would be hard to believe that asymmetry on the particle level could be a factor in determining the handedness of organic molecules.[4]

In discussing the two rival explanations of left-right bias in weak interactions—the space-twist theory and the particle-twist theory—I may have created the impression that the two views are mutually exclusive. This need not be the case. Space may be anisotropic, and this in turn may determine the asymmetric structure of particles and the nature of positive and negative charge. If so, and if the handedness of space is uniform throughout the cosmos, it may be that antiparticles go "against the grain," so to speak, so that it is difficult for them to exist. Antimatter may be completely unstable in our twisted space-time. This would rule out the possibility of antigalaxies. Throughout the entire universe matter would have a uniform handedness.

Most physicists, perhaps only because they have become so accustomed to symmetry on the macroscopic level, find something ugly and unsatisfying about this picture of a universe in which everything twists the same way. One of the most attractive aspects of isotropic space is that it permits the existence of antigalaxies. Note: it only *permits* their existence, it does not guarantee it. For some unknown historical reason, all galaxies may go the same way, just as all amino acids on earth go the same way, even though they could, in theory, go the other way.

At this point speculations about antigalaxies get tangled with theories about the origin of the universe. In the steady state model (now discarded) it was possible to take either view with respect to antigalaxies. According to the model, hydrogen is continually coming into existence to prevent the cosmos from thinning as it expands. If space is anisotropic, this would always be hydrogen, never antihydrogen. If space is isotropic, both types of hydrogen may be arising, but antihydrogen molecules are annihilated by contact with hydrogen already there, just as right-handed amino acids, if they arose in earth's present seas, would be gobbled up by the left-handed organisms already there.

The discovery of background microwave radiation, a pale glow left over from the big bang, ruled out the steady state model, so today's theories about antigalaxies are all in a big bang context. The Swedish astrophysicist Hannes Alfvén, in his book *Worlds-Antiworlds* (W. H. Freeman, 1966) and "Antimatter and Cosmology" *(Scientific American,* April 1967), elaborates on an ingenious metagalactic approach originating with Oscar Klein. In this theory not only may the cosmos contain an equal number of galaxies and antigalaxies intermixed, but even a galaxy may contain an equal mixture of stars and antistars.

There is no way at the moment to be certain that antigalaxies don't exist, or that antimatter isn't flourishing in the cores of galaxies or playing a role in the fantastic production of energy by quasars. For a variety of reasons, however, the consensus among experts is that these possibilities are unlikely. Everywhere in the universe there seems to be only matter. Why?

It is a big, unanswered question. If space is anisotropic, or if asymmetry is built into nature's basic laws, there is no mystery about the absence of antimatter. Physicists, however, have a great fondness for symmetry. Both notions offend their sense of beauty, even though events in the universe seem to go only one way with respect to time—a topic to be covered in the next chapter. One way out is to attribute the universe's lopsidedness to chance events that took place during the first few minutes of the big bang.

If you toss a million pennies into the air it is extremely unlikely that precisely half a million will fall heads, even though no coin is biased. In similar fashion the primeval explosion may have created a slight excess of particles over antiparticles. As the fireball cooled, particles and antiparticles eliminated each other in equal amounts, leaving an excess of particles that turned into our universe.

A rival view is that some sort of repulsive force holds between matter and antimatter. An early suggestion was that the two kinds of matter might repel each other gravitationally. If this were true it would violate general relativity, so physicists were greatly relieved when experiments ruled the conjecture out. There is no evidence yet of any kind of repulsive force, although the notion that the fireball split into two parts, one of matter, one of antimatter, still fascinates many scientists as well as writers of science fiction.

As early as 1956—this before the fall of parity!—Maurice Goldhaber proposed that at the very beginning of time there was a "universon" that split in half to produce a "cosmon" and "anticosmon." The two repelled each other, separating at great speed. Each broke down over billions of years into a universe.

We live in the cosmon. Somewhere out there, perhaps forever beyond our observation, is the vast anticosmon where everything goes the other way. The whole of existence is one gigantic, unthinkable, never-to-be-reunited Mr. Split!

Notes

1. The concept of Mr. Split goes all the way back to Aristophanes' famous speech on love in Plato's *Symposium.* Primeval humans, said the Greek comedy writer, were spherical in shape, with four arms, four legs, and two faces set back-to-back on one neck. There were three sexes: double men, double women, and male and female united in the same body. As punishment for having tried to attack the gods, Zeus split each in half the way one cuts an apple. Love is the desire of

bisected humans to return to their original double form. Heterosexuals are descendants of the male-female type, homosexuals of the all-male and all-female types. "If they continue insolent," said Zeus (and here Mr. Split hops into the picture), "I will split them again and they shall hop about on a single leg."

Freud, in *Beyond the Pleasure Principle,* finds a strong element of truth in Plato's myth and points out that it is older than Plato. The *Upanishads* also ascribe the origin of the first man and woman to a left–right split of an original being. Many great Christian theologians, believing sex to be a consequence of the Fall, have found Plato's myth congenial. "Man is a sick, wounded, disharmonious creature," writes Nikolai Berdyaev, the modern Russian orthodox theologian, in his *Destiny of Man,* "primarily because he is a sexual, *i.e.,* bisected being, and has lost his wholeness and integrity."

A picture of Plato's two-headed spheroid is described as adorning Gargantua's hat in book 1, chapter 8 of Rabelais's *Gargantua and Pantagruel.* In Baum's wonderful fantasy *Sky Island* (Reilly & Britton, 1912), the wicked Boolooroo of the Blues punishes his subjects in pairs by the fiendish method of "patching": the two victims are sliced in half, then the right side of each is glued to the left side of the other.

2. The Kaluza-Klein suggestion of a fifth dimension, perpendicular to the four coordinate axes of space-time, understandably appeals to any Platonist who thinks of this world as a shadow projection of a higher space. We have already mentioned in an earlier note how the concept of a *fourth* dimension was taken over by early spiritualists. Since relativity theory, the "other" world of many occultists has been the *fifth* dimension of Kaluza-Klein. See the appendix on "Five-dimensional Physics" in the first volume of John Gudolphin Bennett's massive three-volume opus *The Dramatic Universe* (Hodder and Stoughton, 1956), for a discussion of the Kaluza-Klein theory and its role in Bennett's brand of occultism. This was written before Bennett became a convert to Subud, surely the funniest of recent religious movements prior to the advents of the Guru Maharaj Ji and the Reverend Sun Moon. For an introduction to Subud, about which I suppress an impulse to write at length, the interested reader is referred to Bennett's *Concerning Subud,* University Books, 1959, and chapter 15 of Steve Allen's autobiography, *Mark It and Strike It.*

3. See Wigner's paper, "Relativistic Invariance and Quantum Phenomena," *Reviews of Modern Physics,* vol. 29, no. 3, July 1957, pp. 255–78, and Landau's "On the Conservation Laws for Weak Interactions," *Nuclear Physics,* vol. 3, 1957, pp. 127–31.

4. Although it is not a popular view, a few scientists have speculated in print on how the asymmetry of weak interactions could induce handedness into primordial amino acids. For example, beta radiation from naturally occurring radioisotopes, falling on a racemic mixture of left and right amino acids, might somehow introduce a left bias. If it turns out that the weak and electromagnetic forces can be unified by a field theory, there could be a slight asymmetry in electromagnetic reactions that would do the trick. Whatever the mechanism, if the asymmetry of basic forces is responsible for biological leftness, then left-life may extend throughout our galaxy, perhaps throughout the entire universe. Right-life would be possible only in an antimatter world where forces go the other way.

25. THE FALL OF TIME INVARIANCE

We know that parity is not conserved; that in our galaxy are forces which produce an asymmetric twist in certain particle interactions. There are strong reasons for believing that in a galaxy of antimatter these twists would go the other way. We know that at least one type of particle, the neutrino, in each of its four inexplicable forms, has an asymmetric structure. Why inversion of charge should also switch left and right, no one knows. There are grave objections to the view that space itself is asymmetrical. There are equally grave difficulties in trying to explain positive and negative charge by left–right reversals of stable asymmetric structures. The view that mirror reflection of matter, in the sense of a pure left-right space inversion, would also reverse charge is nothing more at present than a pious hope.

Yang, in his splendid little book on *Elementary Particles* (Princeton University, 1962), reminds us of Mach's shock when he first witnessed the asymmetric behavior of a magnetic needle in the field surrounding a current of electricity. This mystery vanished, Yang points out, and symmetry was restored when the structure of matter was more deeply understood. Today physicists hope that the mystery of handedness, and the mystery of the electric charge, will similarly vanish with a still deeper understanding of structure. "There are wheels within the wheels," declared Teller in a 1957 speech, "but the real surprise of the whole structure will be that in an unexpected manner and after many more intermediate steps the whole will appear remarkably simple."

In 1964, instead of getting simpler, the situation suddenly became more complicated. This was the year that a group of Princeton University physicists found evidence that another basic symmetry law, time invariance, also appears to be violated in certain weak interactions.

To understand the fantastic implications of this new discovery we must back up a bit in our story and look at the fall of parity in the light of a fundamental symmetry theorem known as the CPT theorem. C stands for electric charge (plus or minus), P for parity (left or right), and T for time (forward or backward). The CPT theorem asserts that, in any natural process, if all three symmetries are reversed, the result is a process that can occur in nature and that is indistinguishable in all other respects from the original one. By "time reversal" a physicist means nothing more than a reversal in the direction a particle (or wave) is moving. A CPT reversal of a glass of milk would mean that all charges would be reversed (making it antimilk), the structure would be mirror-reflected, and every motion would reverse its direction. In more realistic laboratory terms, the CPT theorem says the following. Consider a microevent described by a statement that contains C, P, and T terms, and certain probabilities. Each of the terms has a plus or minus sign in front of it to indicate its handedness, charge, and time direction. Change all three signs but leave the probabilities the same. The new statement will describe a microevent that can occur in nature.

Before the fall of parity, physicists believed that if you altered just one of the signs, the new sentence would still describe something nature could do. Antimilk would be identical with milk except for its charge reversal. Reflected milk would still be milk except that its geometrical structure would go the other way. Time-reversed milk, all its particles moving backward, also would leave the milk unchanged. All the basic laws of physics were believed to be such that, if you reversed time in any mathematical description of an event, you would still have a description of something that could be observed.

"The [basic] laws of nature are indifferent as to a direction of time," was how Arthur Stanley Eddington put it in 1927. "There is no more distinction between past and future than between right and left."

The fall of parity was a surprise, but the overall symmetry of the universe was quickly restored by the discovery that parity reversal is accompanied by charge reversal. Galaxies of antimatter either exist, or are capable of existing, that are identical in every respect with our galaxy except that they are reversed by what physicists call a CP mirror—an imaginary "mirror" that simultaneously reverses both charge and parity.

The 1964 experiments at Princeton involved weak interactions in which CP symmetry was violated. In other words, when both charge and parity were reversed, the resulting event did not duplicate exactly the same event when charge and parity were not reversed. The implications are enormous, for now the only way to preserve CPT symmetry is to assume that time invariance also does not hold. For an explanation of the reasoning behind this assertion, the reader can consult any of three *Scientific American* articles: Eugene P. Wigner, "Violations of Symmetries in Physics" (December 1965); Oliver E. Overseth, "Experiments in Time Reversal" (October 1969); or my own lighter account, "Can Time Go Backward?" (January 1967).

I explained it with three pieces of a small jigsaw puzzle. Imagine a square sliced into three identical rectangles (Figure 67, left). They represent C, P, and T. The square is left-right symmetrical, and so are the rectangles. Turn over (mirror-reverse) any rectangle and it will still fit the square, because its shape has been unchanged. The picture symbolizes the way physicists viewed C, P, and T before 1957.

When parity (P) was found violated, our model indicates this by making the P piece asymmetrical. If we left C and T unchanged, the asymmetry of P would clearly violate CPT symmetry. Why? Because there would be no way to put the three pieces together to make an overall pattern that would remain the same when mirror-reversed.

The middle picture shows how physicists preserved CPT symmetry.

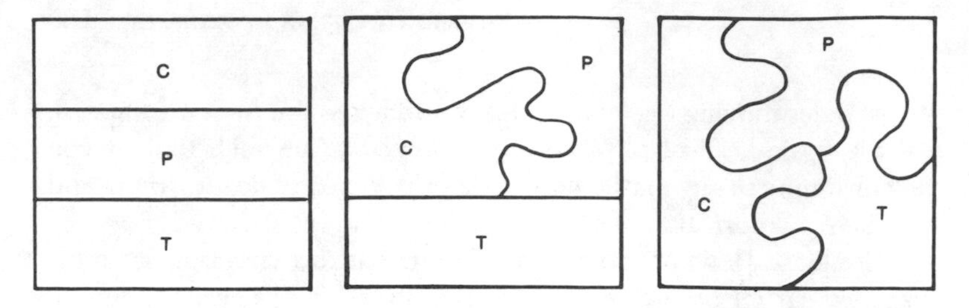

Figure 67. Jigsaw model to explain why CP-symmetry violation implies T-symmetry violation.

Experiments showed that conservation of charge (C) is also sometimes violated. Our model indicates this by giving C an asymmetric shape, but in a way that allows it to fit with the asymmetric P to make a symmetric larger rectangle. A "CP-mirror" reflection, which reverses C and P together, now leaves the combined shape of C and P unchanged. Reverse the CP rectangle and it still fits the symmetrical square.

The third picture symbolizes how physicists reasoned after the 1964 test that the CP shape, too, was not symmetrical. How can we save the symmetry of the CPT square? Only by assuming that the third piece, T, is asymmetrical like the others. In the third picture each piece, as well as each pair of pieces, will violate overall symmetry if turned over. But if all three interlocked pieces are reversed, the square remains a square. CPT symmetry is preserved.

Physicists did not directly observe, nor have they yet, that T has a funny shape. But seeing the funny shape produced by the interlocking of C and P, and assuming that the symmetry of CPT must be preserved, the only way out was to make T asymmetrical. This is why physicists say that a violation of time symmetry is implied by the laboratory evidence, but not directly observed. This is why they still hope and believe that a magic, imaginary "CPT-mirror" reflection would leave all the fundamental processes of nature unchanged.

Consider a stone. Reverse its left–right structure, change all its charges, and reverse all the directions in which anything inside the stone moves. The result is a genuine stone, capable of existing, though not for long if it is in our part of the universe. No one knows, of course, what happens when left–right structure, down to the smallest microdetail, is mirror-reversed. Nor does anyone have the slightest notion of what basic changes occur when charge is reversed. Nevertheless, an *aysmmetry* of time now seems essential if CPT symmetry is to be preserved.

It is important to understand that physicists do not struggle to preserve CPT symmetry merely because they love symmetry. It is true that many physicists find it ugly and unsatisfying to suppose that the world is in some fundamental way lopsided, but the desire to preserve CPT symmetry rests on much more than that. The CPT theorem is so firmly entrenched in the foundations of relativity theory that, if it turns out not to be true, physical theory will be in shambles. "All hell will break loose," was how Abraham Pais once expressed it.

Here is the situation at the moment. No *direct* evidence has been found that a microevent cannot also run backward. CP symmetry *has* been found not to hold in certain interactions. On the assumption that CPT symmetry is conserved, CP violation implies T violation. There are a few theoretical ways to preserve the CP-mirror without combining it with an asymmetrical T, but there is no evidence yet to support any of them. The simplest way is to postulate a superweak "fifth force" that causes the CP anomalies. Experiments have cast strong doubt on the fifth-force hypothesis.

Before exploring some of the wild implications of all this, a few more details about the historic 1964 test may be of interest. The test was made by a group of Princeton physicists, led by Val L. Fitch and James W. Cronin, working at the Brookhaven National Laboratory. In Jeremy Bernstein's words, they used "the same miserable *K* mesons that provoked the theta-tau puzzle out of which parity violation emerged." About once out of every 500 times, the neutral *K*-2 meson (a *K*-2 meson with no charge) decayed into two pi mesons of opposite charge instead of into three pi mesons, as everyone expected. The neutral *K*-2 meson has a CP value of 1. The two pi mesons have a CP value of +1. A CP state of –1 became a CP state of +1, thus violating CP invariance.

In 1966 a team led by Paolo Franzini and his wife Juliet-Lee, also working at Brookhaven, reported a stronger CP violation—this time in an event involving electromagnetism. The neutral eta meson, like the photon, is its own antiparticle. Like the photon, therefore, it would be identical in both matter and antimatter. The short-lived particle (it exists in the laboratory for only a billionth of a billionth of a second) breaks down into three pions, one positive, one negative, and one of zero charge. The Franzinis reported that the + pion had more energy (moved faster in a magnetic field) than its antiparticle. Since this violates CP symmetry, it implies (for the reasons given earlier) a violation of T symmetry.

Alas, before the end of 1966 the eta-decay test was repeated at the European Organization for Nuclear Research (CERN) in Geneva, without finding any violation of CP. The CERN scientists maintained that Franzini's magnetic field, which caused the motions of the charged pions, was uneven. To avoid this bias they continually reversed their field at equal intervals. Final results were negative, and physicists were relieved to return to the view that CP is not violated in electromagnetic interactions. However, in 1967 a Columbia University group repeated the eta-decay test and reported a tiny CP asymmetry, smaller than the Franzinis had found. The statistics are cloudy. At the moment no one is sure whether CP is violated in eta decay or not.

How does the fall of CP, and presumably T, effect the Ozma problem? There are several possibilities. One is that the universe contains no galaxies of antimatter. If this is the case there is, as we have seen, a way by which we and the inhabitants of any planet can arrive at a common understanding of right and left.

A second possibility is that the universe contains galaxies of antimatter in which certain events on the microlevel are time-reversed (motions go the other way) relative to the same events here, but that all macroevents continue to go in the same direction as on earth. In this case, also, the Ozma problem is solved. We have only to ask inhabitants of Planet X to perform a CP-violating experiment. If their description of it tallies exactly with what happens here, we know Planet X is made of matter. If it doesn't, it is made of antimatter. In either case we can discuss experiments that will lead to unambiguous definitions of left and right.

A third possibility—and now we plunge into almost total fantasy—is that the universe may contain galaxies of antimatter in which all events, micro and macro, are moving backward with respect to what Eddington called our arrow of time. Two galaxies can, at least in theory, be time-reversed relative to one another. In each galaxy intelligent creatures would regard their world in the same way that we do ours, as moving forward in time from the past to the future. But they would find events going backward in the *other* galaxy.

Before we can talk meaningfully about such a crazy possibility, and consider whether two galaxies with opposite time arrows could ever become aware of each other, it will be necessary to consider more carefully the nature of time. What is meant by saying that time has a direction? Is there just one arrow of time, or are there many? If there are many, how are they related to one another? Our exploration of left and right is nearing its end as we move from the outer reaches of science into transcendental mysteries.

26. THE ARROWS OF TIME

"... time, dark time, secret time, forever flowing like a river...."

—THOMAS WOLFE,
The Web and the Rock

Time has been described by many metaphors, but none is older or more persistent than the image of time as a river. You cannot step twice in the same river, said Heraclitus, the Greek philosopher who stressed the temporal impermanence of all things, because new waters forever flow around you. You cannot even step into it once, added his pupil Cratylus, because while you step both you and the river are changing into something different. As Ogden Nash put it in his poem "Time Marches On,"

While ladies draw their stockings on,
The ladies they were are up and gone.

In James Joyce's *Finnegans Wake* the great symbol of time is the river Liffey flowing through Dublin, its "hither-and-thithering waters" reaching the sea in the final lines, then returning to "riverrun," the book's first word, to begin again the endless cycle of change.

It is a powerful symbol, but also a confusing one. It is not time that flows but the world. "In what units is the rate of time's flow to be measured?" asked the Australian philosopher J. J. C. Smart. "Seconds per ______?" To say "time moves" is like saying "length extends." As Austin Dobson observed in his poem "The Paradox of Time,"

> Time goes, you say? Ah no!
> Alas, time stays, we go.

Moreover, whereas a fish can swim upriver against the current, we are powerless to move into the past. The changing world seems more like the magic green carpet that carried Ozma across the Deadly Desert (the void of nothingness?), unrolling only at the front, coiling up only at the back, while she journeyed from Oz to Ev, walking always in one direction on the carpet's tiny green region of "now."[1] Why does the magic carpet never roll backward? What is the physical basis for time's strange, undeviating asymmetry?

We have seen how recent laboratory experiments suggest that on the microlevel there are certain weak interactions, perhaps electromagnetic interactions as well, that can go only one way. Thus a time arrow of some sort appears to be built into these events. Aside, however, from these extremely rare anomalies, all the basic laws of physics, including relativity and quantum mechanics, are time-reversible. That is to say, $-t$ can be substituted for t in the statement of any basic law and the law remains as applicable to the world as before; it describes something nature can do.

When we turn to events that occur on the macrolevel, the contrast is at once obvious. An arrow of time is always pointing away from what we call the past, and toward what we call the future. To say that time's arrow always points this way is, of course, a tautology because past and future are defined by the arrow. But that is not the point. The point is that the arrow has a point. There is a difference between its two ends.

The direction of time's arrow is uniform and omnipresent in the workings of our mind. We remember the past. We do not remember the future. I paused after typing that last period. On a sheet of yellow paper were the sentences I had just written — black traces of the past. The rest of the sheet was blank. By now I have typed new sentences. The "now" of that last sentence is now in the past. And now that last "now" has vanished into the past . . . and now . . . Always, for each of us, an uncertain future looms ahead, not yet existing, while the unalterable past is over and done with. It once existed. Now it has utterly disappeared. We know it only by our memories and by the other traces it has left on the present. From these traces we can partially reconstruct the past. Curiously, we do this by inspecting the future, but only at the moment when it is moving into the past. The same mysterious procedure of induction that enables us to guess how nature *will* behave is also used for guessing, with the same varying probabilities, how nature *has* behaved.

It is not only in our consciousness that time's arrow has a fixed orientation. Endless events in the outside world go only one way. Run a motion-picture film backward and who can doubt that it depicts a world we never see? If humans or animals are in the action, the effect is instantly grotesque. If no living things move on the screen, it may not be easy to determine if the film has been reversed until we see certain unmistakable directional clues: a falling leaf, descending snow or rain, waves washing over a beach, or a thousand other one-way things. Why are the fundamental laws of physics, except for certain weak interactions never encountered outside the laboratory, time-reversible, whereas on the macrolevel the universe swarms with events that never go backward?

Before discussing how physicists answer this question, let us try to dispose of a whimsical point of view that is occasionally advanced by a mathematician or philosopher or mystic, though seldom by a physicist. This is the view that only in human consciousness, in the one-way operation of our minds, can a basis be found for time's arrow.[2]

Proponents of this view usually defend it with such opaque language that it is difficult to know precisely what they mean. It seems unlikely they wish to deny that a vast world, external to human minds, exists. If you seriously think a tree has no existence outside your mind, you have no good reason for thinking any other mind exists. There is no logical way to refute you, although surely you must admit that solipsism is not a widespread point of view. Indeed, you must believe that, at the most, there is only one solipsist—yourself.

May I assume, dear reader, that you are not a solipsist—that you grant not only my own existence but also that of a real world *out there*, not made by human minds? If you admit this, you might as well admit that the outside world is structured. The same arguments that support the belief in a tree's existence when no eye is looking at it support equally well the view that the tree has a shape when no eye is looking at it.

There is a sense, of course, in which everything we know about the outside world is what goes on inside our heads. The outside world is always *inferred*, never directly perceived. Information about that world filters into us through our senses, is transmitted in complicated ways along peculiar channels, and is finally interpreted by the brain. In this sense, all that we know about the world is mind dependent. But to say this is to say something obvious and trivial. Knowledge is mind dependent by definition. To know anything is to know it in the mind. If by *sound* you mean a mind's sensation of sound, naturally the tree that falls in a region where no ear can hear it doesn't make a sound. If by *shape* you mean a mind's awareness of shape, naturally a spiral nebula doesn't have a shape. If by *before* and *after* you mean the mind's awareness of before and after, then of course *becoming* is mind dependent.

But what is gained by talking in such an uncommon phenomenological language? Perhaps it gives a perverse sort of pleasure to certain philosophers who can then complain that their critics fail to understand them, and this in turn suggests that their insights are more profound than those of their critics. But to scientists and bartenders, a phenomenological language creates enormous confusion. An astronomer mentions that the Andromeda nebula has a pair of spiral arms. You can't blame him for being annoyed if someone interrupts: "Hold on a minute! Spirality is a mathematical concept of the human mind, not part of nature."

Like ordinary people, and even most philosophers, scientists speak in what Rudolf Carnap liked to call the "object language"—a language that assumes a structured world of things, out there, independent of our brains. Even Bishop Berkeley, who argued so persuasively that nothing exists unless it is perceived, quickly restored the entire outside universe, with all its intricate mathematical properties, by assuming that it is perceived by God.

The point I want to emphasize is that time's arrow is as legitimate a part of the outside world as spatial relations. The thing the arrow stands for, the one-wayness of events, is "out there" in the same sense that large and small, hot and cold, fast and slow, light and dark, left and right, and all the other structural relations of the world are out there. Did dinosaurs have grandchildren? If so, dinosaur grandmothers were older than their grandchildren. Even the "now" was out there in the sense that, for every dinosaur, there was a "now" when it was born.

Perhaps the subjectivist counters with, "Yes, but there were dinosaur minds to perceive the passage of time." Okay, then we go further back in time. Did trilobites perceive the passage of time? Did the one-celled forms of life in the primeval oceans? Let us go back still further. Mars is older than its craters. The sun is older than Mars. The Milky Way galaxy is older than the sun. If the big bang theory is true, the universe began in a monstrous explosion billions of years ago and has been expanding ever since. Our *awareness* of that expansion is of course mind dependent, but why bother to assert the obvious?

The arrow of time inside our head clearly points the same way as the arrow of time on the outside. Mars's surface kept a record of the past. Our brain cells keep a record of the past. Why do the two arrows correspond in direction? Everyone except a few subjectivists would answer, in the objective language of science, that it is because our brains are made of the same stuff as the universe, and its particles dance to the same laws. Our awareness of time depends on memory, and memory is just a complicated kind of footprint. Surely it requires a strange sort of narcissism to suppose that our feeble little brain imposes its arrow of time on the cosmos rather than the other way around.

If you agree that becoming is part of nature, independent of our minds, then we are back to our earlier question. Since the fundamental laws of physics (except for the rare anomalies we have noted) are time-reversible, what is it that keeps nature always moving in the same direction? Why do so many events in nature go only one way?

Part of the answer, perhaps all the answer, lies deep within the laws of probability. Certain events go only one way not because they can't go the other way but because it is extremely unlikely that they go backward. To grasp what is meant by this, I know of no better method than to perform a few simple experiments with a deck of playing cards.

Consider a "deck" of cards that consists of only the ace of spades. "Shuffle" this deck as much as you like, then examine the "order" of the cards from top down. The probability it is still the ace of spades is

1 (certain). Now repeat the same procedure with the ace and deuce of spades. The situation has already started to get mysterious. After sufficient shuffling, to randomize the order of this two-card packet, the probability that the sequence, from top down, is ace–deuce is 1/2. The probability it is deuce–ace is also 1/2.

Try it with the ace, deuce, three, and four of spades, starting with the cards in that order from top down. Four cards can be arranged in $4! = 1 \times 2 \times 3 \times 4 = 24$ ways. The probability is 1/24 that after thorough shuffling you will find the cards back in 1–2–3–4 order. If you keep shuffling and periodically examining the packet, you will discover that, in the long run, about one out of every twenty-four examinations will show the packet to be in 1–2–3–4 order (or any other specified order).

The number of ways n objects can be arranged is $n!$ or factorial n. Factorials increase in size at a fantastic rate. With a standard deck of fifty-two cards, the probability that random shuffling will return the deck to its original order is 1/52! or 1 over a number that is 8 followed by 67 digits. If you open a new deck of cards, write down the way they are ordered, then shuffle the cards thoroughly, it is a safe bet that the original order becomes hopelessly lost. Nothing in the basic laws of physics prevents the deck from returning to its initial state, only the laws of chance. If the deck is shuffled long enough, say by a shuffling machine that runs for a few million years, the original order might be restored. Indeed, there is a famous theorem of Poincaré's which asserts that, given a sufficient amount of time, the initial order is certain to return as many times as you care to specify. If the shuffling goes on forever, it will return an infinite number of times.

Whenever a large number of things interact with one another in random ways, probability introduces a one-wayness in time. A popular way to explain this is to describe what happens at the start of a pool game. Imagine a motion picture of the break by a cue ball of the fifteen numbered balls that are closely packed into an equilateral triangle. The balls scatter hither and thither, and the eight ball, say, drops into a side pocket and rolls down. Show this movie backward and everyone instantly recognizes that the film has been reversed. No pool player, playing after the break, has ever seen the balls roll around on the green cloth and come together again to form the triangle. The essential point is that no law of physics prevents this. It fails to happen only because it is extremely improbable.

What about the eight ball that dropped into a pocket? Surely, you might contend, physical laws forbid it from rolling back up the incline and leaping out of the pocket to join the other balls. No, this event also is prevented only by probability. Now you must think of molecules as behaving like billions of microballs bouncing around in space. Imagine a flask filled with a certain gas. Open the flask and in a short time the gas molecules have shuffled themselves evenly around the room. They do not shuffle themselves back into the flask for the same reason that fifty-two cards don't shuffle back to an original sequence. It is too unlikely. If you imagine every gas molecule in the room suddenly reversing its direction, the molecules *would* go back into the flask. Gas diffusion is time-reversible in principle. It doesn't happen in practice because of statistical probability laws.

Let us see how this applies to the eight ball. After all the balls have stopped moving, imagine that the motion of every molecule that played a role in this event is reversed. At the spot under the table where the eight ball came to rest, the molecules that carried off the heat and shock of impact would all converge to create a small explosion. This explosion would send the ball rolling up the incline. Along the way, molecules that carried off the heat of friction would move toward the ball and boost it along its upward path. All the other balls would be set in motion in a similar fashion. The eight ball would pop out of the side pocket and the balls would roll about until they finally converged to form the initial triangle. The impact of this convergence would shoot the cue ball back toward the tip of the cue.

A motion picture of the behavior of any individual molecule in this event would show nothing unusual. No law of physics would be seen violated. But when we consider the billions upon billions of "hithering and thithering" molecules involved in the event, the probability that they would all move in this way, to time-reverse the event, is so close to zero that if we saw such a thing happen we would think we had seen a miracle.

Because gravity is a one-way force, always attracting and never repelling, one might suppose that the motions of bodies under the influence of gravity could not be time-reversed without violating basic laws. Such is not the case. Reverse the directions of the planets and they would swing around the sun in the same orbits. It is this reversibility that permits astronomers to calculate exactly when eclipses occurred in the past.

What about the collisions of objects drawn together by gravity—the fall of a meteorite, for instance? Surely *this* is not time-reversible. But it is! When a large meteorite strikes the earth, there is an explosion. Billions of molecules scatter hither and thither. Reverse the directions of every molecule and their impact at one spot would provide just the right amount of energy to send the meteorite back into orbit. An egg drops off a wall, like Humpty Dumpty, and splatters on the ground. Reverse all the molecules in this event and the egg shuffles itself back together and hops to the top of the wall. No basic laws would be violated by such an event, only statistical laws.

It is here, then, in the laws of probability, that most physicists find the ultimate basis for time's arrow. Probability explains such apparently irreversible processes as the mixing of coffee and cream, the breaking of a window by a stone, and all the other familiar one-way-only events in which large numbers of molecules are involved. It explains the second law of thermodynamics, which says that heat always moves from hotter to cooler regions. It explains why shuffling destroys the order of a deck of cards.

"Without any mystic appeal to consciousness," Eddington declared (in the lecture in which he introduced the term *time's arrow*), "it is possible to find a direction of time. . . . Let us draw an arrow arbitrarily. If as we follow the arrow we find more and more of the random element in the state of the world, then the arrow is pointing towards the future; if the random element decreases the arrow points towards the past. That is the only distinction known to physics."

Eddington was well aware, of course, of another large class of events that always go one way: events in which energy radiates from a center. A pebble dropped into a pond creates expanding circular ripples. We never see circular ripples contract in a pond, converge on a pebble, and propel it out of the water. Here again it is only because such an event is improbable. In principle, given the right set of initial conditions, such an event would actually occur. Indeed, in a laboratory tank it is quite easy to produce water waves that form circular ripples which contract toward a point. In nature, of course, we cannot conceive of events at the rim of a lake that would create such a phenomenon, quite aside from all the molecular motions in the water that would have to be reversed to carry a sunken stone upward to the precise spot where the converging ripples would send it out of the water.

The same near-zero probability applies to a thousand other radiative events that can be reversed in theory but not in fact. If an electron, a proton, and a neutron were all shot from outer space with such accurate aim that all three struck the same nucleus of an atom, they would create a neutron. Time-reversed beta-decay would be observed. We never observe it, not because nature forbids it, but because the initial conditions for it are too improbable.

A star's radiation illustrates the same thing on a larger scale. We never see the energy coming from all directions to converge on a star with backward-running nuclear reactions that would make the star an energy sink instead of a source. Here again, there are no fundamental laws of physics (including relativity laws and the laws of quantum mechanics) to forbid such an event. It is only the improbability of initial conditions that forbids it. One would have to assume that God or the gods, in some higher continuum, started all the waves at the rim of the universe. Without such improbable "boundary conditions" at the rim of things, there is no way to get the backward radiation process started.

On the largest scale of all, the steady expansion of the universe is a radiation from a center where the primeval explosion occurred. Once more, there is no reason why all the matter in the universe couldn't go the other way. If the directions of all the receding galaxies were reversed, the red shift would become a blue shift, and the shrinking cosmos would violate no known basic laws. As we shall see in chapter 28, many cosmologists believe that some day the universe actually will stop expanding and start going the other way.

For Eddington, all such radiative processes, from concentric ripples on a lake to the expanding universe, are simply other instances of movement toward disorder. At first the ripples near the stone are highly ordered. As they move outward they become progressively less so, until finally they vanish. Light radiating from a sun becomes more disordered as it is influenced by other astronomical bodies and by the curves of space-time. If space-time is closed like the surface of a sphere (as Eddington believed it to be), eventually a star's radiation will become completely chaotic. The universe on the whole, Eddington argued, is steadily expanding into a state of maximum disorder. In any case, he was persuaded that all events in the physical universe, which go only one way in time, owe their direction to the laws of probability.

A fascinating question now arises. If time's macroscopic arrow is defined by probability as the direction in which events move toward increasing randomness, how did the universe ever manage to get itself into an initial highly ordered state? If the universe is running down, what wound it up? This is the next chapter's topic.

Notes

1. From a mental point of view, the "now" is not the physicist's instant, but a fuzzy region of a second or two during which the brain retains a strong image of an immediate event, like the retinal image that gives the illusion of continuous motion to a film, or the mind's ability to hear a chord when notes are played in sequence. The point has been stressed by William James, Josiah Royce, and many other philosophers and psychologists.

2. The classic statement of this position, often quoted, is by the mathematician Hermann Weyl: "The objective world simply *is*, it does not *happen*. Only to the gaze of my consciousness, crawling upward along the life-line [the world-line of relativity theory] of my body, does a section of this world come to life as a fleeting image in space which continually changes in time."

Some philosophers have defended this view, and it has held a strong fascination for a few scientists as well. In recent years its most vocal champion has been Adolf Grünbaum. Grünbaum grants that time is anisotropic in an objective sense, but he argues that "becoming," with its notion of "now," is wholly mind-dependent. Karl Popper, one of the many philosophers of science with whom Grünbaum has clashed over this matter, has expressed a wonderment (which I share) that Grünbaum, a firm realist, suddenly talks in the language of idealism whenever he discusses the arrow of time. (See *The Philosophy of Karl Popper*, edited by Paul Arthur Schilpp, page 1,141.)

I like to think that Grünbaum's perversity here is entirely linguistic, rather than a fundamental disagreement with such realists as Popper and Bertrand Russell. It seems to me that Grünbaum gives the game away when he writes (at the close of his paper on "The Meaning of Time," in *Basic Issues in the Philosophy of Time*, edited by Eugene Freeman and Wilfrid Sellars). "But in characterizing becoming as mind-dependent, I allow fully that the mental events on which it depends themselves require a biochemical physical base or possibly a physical base involving cybernetic hardware." If Grünbaum believes that, why all the fuss about the language of realism?

27. ENTROPY

Entropy has a precise technical definition in both thermodynamic theory and information theory, but for our purposes it will be sufficient to think of it in a rough way as a measure of disorder—the absence of pattern. The "information content" of a system, again in a rough way, is a measure of order. The two measures vary inversely. If the entropy of a system goes up, its information content goes down, and vice versa.

Both properties are easily demonstrated with our deck of cards. Let us arrange it so that black and red cards alternate, a black card on top. So far as the colors go, we have complete information about all the cards, but their suits and values may be random. Compare this with another deck in which only the values are random. For example, the suits are arranged, top down, in repeating cycles of spades, hearts, clubs, and diamonds. The second deck has lower entropy and higher information content than the first one.

Let us shuffle either deck. As the cards are mixed, the deck's entropy rises and its information content goes down. (Riffle shuffles, by the way, are surprisingly inefficient. It takes a great many such shuffles to destroy completely the order of a highly ordered deck.) When the cards have been sufficiently shuffled, the deck reaches a state of maximum entropy and minimum information content. It then corresponds roughly to what in thermodynamics is called a state of thermal equilibrium.

Ludwig Boltzmann, the nineteenth-century Austrian physicist who founded statistical thermodynamics, was one of the first scientists to speculate deeply about how statistical laws might be used to explain why our universe is as strongly patterned (low in entropy) as it is. This was not a question that had troubled earlier scientists. Galileo, Kepler, and Newton would have agreed on a simple answer. God is a great mathematician, and "the firmament sheweth his handywork." Boltzmann lived at a time when scientists and philosophers were beginning to think about cosmology in evolutionary terms. Is it possible, Boltzmann asked himself, that the universe was initially a vast chaotic sea of randomly moving particles? Did the laws of entropy dictate that somewhere in this formless sea a universe as intricately patterned as ours would arise naturally?

The starting point of Boltzmann's daring vision was a system of gas molecules moving about randomly in a closed container. We must idealize this model by assuming it to be completely isolated from the rest of the world. The gas molecules race hither and thither, bouncing off the walls and off each other. All their motions are time-reversible. If we could take a micromovie of any part of the gas and run the film backward, what we would see would be indistinguishable from what we would see if the film ran forward. The gas is in thermal equilibrium—a condition of maximum entropy.

If nothing existed in the universe except this container and its gas, could we say that the system possessed an arrow of time? No, said Boltzmann, we could not. Time may be there in a sense, because we cannot conceive of motion without time, but there is no way to orient an arrow of time to distinguish one direction from the other. The film looks the same either way. It is symmetrical, directionless, arrowless time. At this point, to avoid hopeless paradox, we must think carefully about the role of the observer.

Suppose that you, with your own psychological arrow of time, are watching the molecules in this container. You have the ability to see the behavior of each individual molecule. The system is no longer arrowless, because you are now imposing upon it *your* arrow of time.

You see each molecule moving from *your* past to *your* future. If every molecule suddenly stopped and reversed direction, you would observe thissss at once. You would not, of course, say that time had reversed. You would say that the molecules had inexplicably reversed their motions without interrupting the steady forward flow of events.

The container when observed is obviously no longer an isolated system. It is interacting with you, a complicated molecular system with a well-defined time direction. Try, now, to think of the container as the only existing thing. There are no observers, not even gods. In some sense, perhaps, time is still there, but there is no way to define a direction for it. To say that time reversal occurred in such a system would be as meaningless as saying that the gas turned upside down or mirror-reversed.

The situation is much like one we encountered earlier in connection with handedness. The letters OUT painted on a glass door spell neither OUT nor TUO unless there is an observer present to stand on one side or the other of the glass. Imagine nothing existing in the universe except OUT, suspended in the void. It has handedness, but there is no way to define the kind of handedness. To say that OUT mirror-reversed itself is to say nothing.

Let us take a closer look at what is going on inside our idealized container. Its molecules are dancing randomly. The overall state is one of maximum entropy. As we watch, however, we see a wondrous and delightful thing happening. Here and there little pockets are forming where entropy momentarily decreases, then increases again. At first we will notice myriads of these tiny, short-lived areas where the molecules bunch together more than usual. If we watch long enough we will see larger, longer-lived patches of order. If we watch an enormously long time, billions upon billions of years, eventually we will see the molecules form any pattern we care to define, such as all gathering in one corner of the container, or momentarily arranging themselves to spell the word OUT. Within any such patch of order we can assign an arrow of time to show whether the patch is moving toward increasing or decreasing entropy; more precisely, to show which way we guess the patch is moving. Unfortunately, at any one moment a patch is just as likely to be going one way as the other. There is no way that we can give a preferred time direction to the system as a whole. It fluctuates in spots, and even its overall entropy fluctuates, but on the whole it remains in a state of arrowless thermal equilibrium.

These fluctuations can be beautifully modeled with our deck of cards. Shuffle the deck thoroughly, then spread it face up on the table and examine its sequence. You will find it filled with little patches of order. Here is a run of five red cards, there are two jacks together, here is a sequence of 6–7–8, and so on. We, of course, are defining what we mean by *pattern*, but it doesn't matter. If we decided that a run of king–7–10 was a pattern, then we would find *that* pattern occasionally turning up. Shuffle the deck some more. Whatever patterns ou found before will blend back into the overall chaos of the deck, and new patches of order will form. The longer you keep shuffling, the more likely you will find surprisingly large patterns. In the long run you can expect to find any sequence of cards you care to define, including sequences that specify the position of every card in the deck. Moreover, any defined sequence will return infinitely many times if the shuffling continues forever. If we devise a cipher system so that the cards encode letters, an infinite shuffling of the deck will spell out all the plays of Shakespeare, and do this infinitely many times!

Boltzmann's great vision was of a cosmos of stupendous size, perhaps infinite in space and time, made of countless particles in thermal equilibrium. Would it not be inevitable, he reasoned, that here and there pockets would form in which entropy would for a short time (the "short" time could be billions of years) decrease? Our universe, a flyspeck portion of the infinite sea of chaos, is perhaps such a region. At some time in the past, eons ago, entropy just happened to decrease enough at one spot to form the ordered universe in which we find ourselves. The event is so unlikely that it seems to us a miracle, but we must remember that an infinite time may have passed (whatever that means!) before our universe fluctuated itself into existence. We are the actors in a play that has been accidentally produced by the infinite shuffling of cosmic cards. Strictly speaking, we cannot talk about the chaotic sea as moving in any time direction. All we can say is that within our little play, our momentary pocket of low entropy, time has acquired an arrow. As our universe now fades back into thermal chaos, its arrow of time points in the direction of increasing entropy.

This is not the place to go further into the tangled difficulties and paradoxes that arise when Boltzmann's vision is elaborated. No physicist today takes it seriously, although occasionally someone tries to revive it in terms of quantum fluctuations in empty space or the "many worlds" interpretation of quantum mechanics. It all works very well for idealized closed systems of gas molecules, but the universe itself is much more interesting. For one thing, if the universe behaved like a fluctuating gas in thermal equilibrium, we would expect, as we look into distant regions of the universe, to find portions of it in greater disorder than the portion we occupy. Instead, we see that it is as well ordered as our own region.

However, all is not lost in finding a way to preserve Boltzmann's central notion that entropy is the principal foundation for the arrow of macroscopic time. The key to everything is the big bang, and it is hard to imagine that Boltzmann would not have been delighted had he lived to see the evidence for this explosive origin of the universe become overwhelming. No longer is it necessary to grope for explanations of low entropy in the eternal fluctuations of randomly moving particles. The big bang does the job for us in just a few minutes!

It all begins, of course, in mystery. Who knows what caused the primeval explosion or exactly what the material was that exploded? If Boltzmann were alive he would probably attribute it to a random event that occurred in a sea of quarks, shifting about in a state of maximum entropy. The quark soup may have contained some sort of time, or it may be meaningless to speak of time before the explosion. Anyway, a moment later, when particles and energy were radiating outward with unimaginable speed, and the fireball's temperature rocketed downward, entropy suddenly dropped and a universe of beautiful macroorder was born. As it was born, two enormous arrows of time were stamped upon it: the direction of the cosmic expansion, and the arrow of entropy.

The two arrows are not the same, and it seems to me useless to argue about which one is the more basic. The universe owes its great drop in entropy to the big bang, so in that sense we can say that the explosion, and the resulting expansion, caused and now preserves the entropy arrow. This is a currently fashionable way of talking about the universe. See, for example, David Layzer's article, "The Arrow of Time" (*Scientific American,* December 1975), in which he shows how the expansion, combined with quantum mechanics and the inverse relationship of entropy to information, together produce and maintain the thermodynamic arrow.

On the other hand, it is hard to imagine how the expanding universe could have begun its long journey from order to disorder unless laws of probability did not, in some unfathomable sense, predate the bang. I suppose one could argue that laws of probability were also created by the bang, but now we are in such a metaphysical region that we can't pursue the matter further without going into age-old debates about the foundations of mathematics.

However one talks about it, the universe is an obviously rich mixture of vast, overall movements toward chaos, with small patches where things are moving the other way. Layzer uses the term *historical arrow* for those processes in which order is increasing. The formation of matter, moving in an orderly fashion outward from the site of the big bang, was the first gigantic instance of an event stamped with the historical arrow. The evolution of stars and planets is later examples. Finally, at least on one planet, the energy radiating from a highly ordered sun allowed the rise and proliferation of life, the most highly patterned thing we know.

To understand the history of these patches of increasing order, moving against the grain of increasing entropy, it is useful to introduce what Hans Reichenbach, an eminent German philosopher of science, called branch systems. These are quasi-isolated systems in which entropy is either increasing or decreasing rapidly, sometimes very rapidly, while the universe as a whole drifts toward chaos. P. C. W. Davies, in his valuable little book *Space and Time in the Modern Universe* (1977), describes a typical hierarchy of branch systems. It starts with someone dropping an ice cube into a glass of hot water. The entropy of the water immediately rises. This is *not* a natural Boltzmann-type fluctuation. It is a strongly time-asymmetric process created by our sudden action of creating a branch system (cube and water) that was not there before.

Consider now the cube of ice, a strongly ordered crystal. It got to be what it is from another branch system, a refrigerator. The refrigerator seems to violate the second law of thermodynamics by moving heat out of a cold region and into the warmer region of a room. This causes the water to freeze and form the ice cube. The refrigerator is able to do this because it, too, is influenced by another branch system, a heat pump. The pump operates on electrical energy that comes from still another branch system, a dynamo. The dynamo in turn uses energy from, say, the burning of oil. The highly ordered energy locked in oil takes us back to millions of ancient branch systems, the animals and plants whose highly patterned bodies formed the oil. These living things were intricate systems of "negentropy" (negative entropy) whose arrows went the way they did because they fed on the increasing entropy of our highly ordered sun.

The earth's surface swarms with billions of branch systems stamped with an entropy arrow that points in one direction or the other. Almost all of them operate on energy that ultimately comes from the sun. Volcanoes and earthquakes are exceptions, but wind and the motion of water are sun-dependent energy sources because it is the sun that keeps the atmosphere in a perpetual state of entropy unbalance. A city such as London represents an enormous growth of order and information. London could not have evolved except as the result of vast movements toward disorder in the world outside it. The city's millions of branch systems, living and nonliving, operate on entropy disequilibriums produced by chains of systems that ultimately link to the sun.

A house is built. That is a slow growth of order. A bomb obliterates it. That is a fast growth of disorder. Neither event is a Boltzmann-like fluctuation. Both are the result of strong interference by outside agencies. In the long run, the second law of thermodynamics prevails. The cosmos cannot, as a whole, lose entropy. Inside it are these peculiar pockets where things are winding up. But the great majority of branch systems are running down, which means that the universe on the whole is running down.

We must not become linguistically confused by the fact that the historical arrow and the entropy arrow point opposite ways with respect to order. It is easy to distinguish between the two kinds of movements. Wind can blow down a house of cards; it can't blow the house back into existence. A flower never grows into the ground. The falling house of cards is a movement toward disorder. The growing flower is a movement toward order. Both events are unidirectional. Both define the *same* direction of time in the macroworld. Henceforth we will ignore the distinction between the two kinds of movement and employ the term *entropy arrow* for the uniform flow of macroevents, wherever entropy is involved, from past to future.

As for the universe itself, is it really running down? Cosmologists don't know for sure, even though they like to talk sometimes as if they do. If space-time is open (not curved back on itself, as in Einstein's original model), the expansion will continue forever. The universe will eventually reach a burned-out, inert state that Sir James Jeans called the heat death—by which he meant death from freezing. All the stars will become cinders—either black dwarfs or neutron stars or black holes. Their old radiation will just keep on going forever into the bleak nowhere of outer space.

If space-time is closed (at the moment, this seems unlikely), the expansion will eventually halt and the universe will go into a contracting phase. What happens next, and how the contraction will affect stars, planets, and life, is far from clear. Presumably, overall entropy will start decreasing as the cosmos moves toward its inevitable implosion into a black hole. We will consider this in the next chapter.

Let us summarize. Within our crazy universe, as we find it, there are at least five arrows of time. Physicists do not yet know how they are interrelated. The preferred time direction on the microlevel, in certain weak interactions involving *K*-mesons, is still a total mystery. It may have no connection with the macroscopic arrows, just as the handedness of particles seems to have no connection with the handedness of molecules, and the handedness of molecules in turn has no bearing on the bilateral symmetry of a tiger.

On the macrolevel are four arrows. First, there is the entropy arrow we have been discussing. Second, there is the arrow defined by events radiating from a center like expanding circular ripples on a pond or energy radiating from a star. Both these arrows derive from probability laws—the entropy arrow from the statistical laws of thermodynamics, the radiative arrow from the probability of initial or boundary conditions. Third, there is the expansion of the universe. Fourth, there is the psychological arrow of consciousness.

Can one or more of these five arrows be reversed without affecting the other arrows? Can *all* of them be reversed? In recent years there has been considerable speculation, not only by science-fiction writers and philosophers, but also by top cosmologists, about the existence of universes in which most or even all of the five arrows go the other way with respect to our own arrows. These time-reversed fantasy worlds will be investigated in the next chapter.

28. TIME–REVERSED WORLDS

We have seen how the direction of time can be defined by five arrows. Let us put aside the question of how those arrows relate to one another and consider the following question: is it meaningful to talk about the existence of another universe, exactly like our own with respect to all basic laws, but with all five of its time arrows pointing the opposite way from ours?

The CPT theorem suggests that such a world would be made of antimatter. There is strong evidence that antimatter (charge-reversed matter) has a handedness that is the reverse of matter. As we have learned, the violation of CP symmetry implies T asymmetry. It would be aesthetically pleasing to theoreticians if a universe could exist in which C, P, and T were all reversed. As a thought experiment, let's assume that a reversal of T on the particle level is combined with a reversal of the other four time arrows. Is it possible that somewhere out there, in another space-time continuum, there is an antimatter universe in which structures not only go the other way in space, but also go the other way (in all respects) in time?

Two worlds with opposite time arrows are analogous to two worlds that are mirror images of each other. If we eliminate the role of an outside observer with his own sense of left and right, then all we can say is that each world is a mirror reversal of the other. The same is true of time-reversed worlds. In each universe intelligent beings are living "forward" in their time. To say that time in one universe is "backward" means nothing more than that events in that universe go the other way relative to events in the first universe.

This notion of two worlds with contrary arrows of time goes back to Boltzmann. There seems to be nothing logically contradictory about it, although it leads to many bizarre results. For example, no two-way communication is possible between intelligent minds in the two time-reversed worlds. To see why, suppose that some kind of communication channel is established between person A in one universe and person Z in a time-reversed world. A sends a message to Z. Z manages to decode it and send a reply back to A. From A's point of view, Z is moving into Z's past. Z can't reply because he hasn't yet got the message. From Z's point of view, any reply that he makes will arrive in A's past *before* A sent the original message! From either point of view, logical contradictions arise if the possibility of a reply is assumed. The situation is similar to those paradoxes that arise in science-fiction stories when a person travels into the past and murders himself as a child.

Communication, therefore, seems to be ruled out by logic. What about observation? It is easy to see a mirror-reversed world—just look into a mirror! But seeing a time-reversed world poses difficulties. For one thing, light, instead of radiating from the other world, would be going toward it. If observation involved electromagnetic radiation, each world would be totally invisible to the other. Let us pull out all stops and suppose that some day we will discover a type of radiation that can be directed toward a time-reversed world and which will bounce back to us without interfering in any way with the other world's history. By the use of this strange radiation we can "observe" what is happening in the other world, although we cannot use it for any kind of communication. We will, of course, see the other world as moving backward in time. They in turn can use the same technique to see us moving backward in time.

No one has the slightest idea of what such a radiation would be like, but there seems to be no logical contradiction that follows from assuming it to exist. Oddly enough, the assumption does not even presuppose a deterministic view of history—the view that, given the state of the universe at any one moment, the entire future of the universe is uniquely determined. When A probes the state of Z's universe, all he can observe is that universe going back into its previous states. Similarly Z, probing A's universe, sees that universe going backward. The past, as everyone agrees, is fixed for all eternity. Because neither A nor Z can probe the other world's future, both futures remain indeterminate. Seeing into the past of another universe has no more effect on the determinism-indeterminism controversy than seeing a motion picture of some past historical event in human history.

Gods in higher space-times, observing two universes with opposite time arrows, are of no help in thought experiments intended to settle the controversy. Even if they see the entire history of both universes, in one blinding instant of hypertime, it does not preclude the possibility that each universe, as it unrolled in its own time, did not have branch points that were undetermined at each moment of branching. Indeed, this is precisely how many great theologians of all religions have reconciled the seemingly contradictory notions of free will and predestination. The ancient debate between determinists and indeterminists appears to be unaffected by the notion of time-reversed worlds.

Frank Russell Stannard, a British physicist writing on "Symmetry of the Time Axis" (*Nature*, August 13, 1966), suggested (not too seriously) that two time-reversed worlds might occupy the same volume of space-time by interpenetrating each other but not interacting in any way, like a pair of checker players playing one game on the black squares while another pair play a different game on the red squares. He called the "other" world faustian because Faust, in Goethe's poem, was permitted by Mephistopheles to go back in time. In Stannard's vision the faustian world is all around us, going the other way in time, forever cut off from our observation.

J. A. Lindon, my favorite writer of comic verse, was moved by Stannard's vision to compose the following poem:

NOT *THAT* WAY!

So I slipped through the doorway that said DR. STANNARD—
  Oh, mental kaleidoscope! Everything swirled!
Here all that occurred seemed outlandishly mannered,
  For *time goes reverse* in the *faustian* world!

I saw Mr. Crankylank backwardly biking
  (But safe) through the traffic that rumbled so near;
Off home to his lunch (which had been to his liking),
  The train he had missed coming in, by the rear.

I glimpsed Eddie Champer unchewing his bacon
Before turning in for his morn-to-night's rest;
His wife then unfrizzled it fit to be taken
And sold to the grocer marked BACON BACK BEST.

I bowed at the "hundredth" of old Lady Brinker,
Who died last December (though buried before);
With a chuckle this blinking fat-winking hard-drinker
Said she'd taste mother's milk in one century more!

Men showed me the prison where Bill the Bank Robber
Was serving his sentence of retrograde time;
Come seventeen years, he might doff prison clobber,
Be led out and left decommitting his crime.

I spied on Lou Cleanbody bathing. (The struggle
With my better nature defeated me, chaps!)
She soaked up the scum that emerged from the plug-hole,
Got out, and fresh water poured into the taps.

I viewed an unbombing. Debris reassembled
Itself into buildings, life came out of doom.
Smoke flashed into bombs, which flew up (as I trembled)
And hooked under planes in a tail-upward zoom.

I heard of unsabotage: works had been spannered;
The spanners flipped out, broken cogs then OK.
Here I left by the doorway that said DR. STANNARD,
And found my watch going the usual way.

But I wrote a great book on it. No one will quote it,
Though I'd learned it all thoroughly, knew it right through.
I began at the end and forgot as I wrote it,
Unscribbled the title, know no more than you!

If Stannard's vision seems fantastic, consider the following notion, which goes all the way back to Plato. Suppose that the expanding universe reaches a point at which gravitational forces halt the outward drift and the universe starts contracting. Perhaps at the extreme limit of the expansion our world will enter a space-time singularity—a point at which the equations of physics no longer apply—then when it starts to contract, all the time arrows will spin around and point the other way. The universe, in brief, will turn into a time-reversed world of antimatter.

In recent years the English astrophysicist Thomas Gold has seriously proposed just such a cosmological model, but first let us see how Plato describes it in his dialogue *The Statesman:*

STRANGER:

Listen, then. There is a time when God himself guides and helps to roll the world in its course; and there is a time, on the completion of a certain cycle, when he lets go, and the world being a living creature, and having originally received intelligence from its author and creator, turns about and by an inherent necessity revolves in the opposite direction.

SOCRATES:

Why is that?

STRANGER:

Why, because only the most divine things of all remain ever unchanged and the same, and body is not included in this class. Heaven and the universe, as we have termed them, although they have been endowed by the Creator with many glories, partake of a bodily nature, and therefore cannot be entirely free from perturbation. But their motion is, as far as possible, single and in the same place, and of the same kind; and is therefore only subject to a reversal, which is the least alteration possible. For the lord of all moving things is alone able to move of himself; and to think that he moves them at one time in one direction and at another time in another is blasphemy. Hence we must not say that the world is either self-moved always, or all made to go round by God in two opposite courses; or that two Gods, having opposite purposes, make it move round. But as I have already said (and this is the only remaining alternative) the world is guided at one time by an external power which is divine and receives fresh life and immortality from the renewing hand of the Creator, and again, when let go, moves spontaneously, being set free at such a time as to have, during infinite cycles of years, a reverse movement: this is due to its perfect balance, to its vast size, and to the fact that it turns on the smallest pivot.

SOCRATES:

Your account of the world seems to be very reasonable indeed.

STRANGER:

Let us now reflect and try to gather from what has been said the nature of the phenomenon which we affirmed to be the cause of all these wonders. It is this.

SOCRATES:

What?

STRANGER:

The reversal which takes place from time to time of the motion of the universe.

SOCRATES:

How is that the cause?

STRANGER:

Of all changes of the heavenly motions, we may consider this to be the greatest and most complete.

SOCRATES:

I should imagine so.

STRANGER:

And it may be supposed to result in the greatest changes to the human beings who are the inhabitants of the world at the time.

SOCRATES:

Such changes would naturally occur.

STRANGER:

And animals, as we know, survive with difficulty great and serious changes of many different kinds when they come upon them at once.

SOCRATES:

Very true.

STRANGER:

Hence there necessarily occurs a great destruction of them, which extends also to the life of man; few survivors of the race are left, and those who remain become the subjects of several novel and remarkable phenomena, and of one in particular, which takes place at the time when the transition is made to the cycle opposite to that in which we are now living.

SOCRATES:

What is it?

STRANGER:

The life of all animals first came to a standstill, and the mortal nature ceased to be or look older, and was then reversed and grew young and delicate; the white locks of the aged darkened again, and the cheeks of the bearded man became smooth, and recovered their former bloom; the bodies of youths in their prime grew softer and smaller, continually by day and night returning and becoming assimilated to the nature of a newly-born child in mind as well as body; in the succeeding stage they wasted away and wholly disappeared. And the bodies of those who died by violence at that time quickly passed through the like changes, and in a few days were no more seen.

SOCRATES:

Then how, Stranger, were the animals created in those days; and in what way were they begotten of one another?

STRANGER:

It is evident, Socrates, that there was no such thing in the then order of nature as the procreation of animals from one another; the earth-born race, of which we hear in story, was the one which existed in those days—they rose again from the ground; and of this tradition, which is now-a-days often unduly discredited, our ancestors, who were nearest in point of time to the end of the last period and came into being at the beginning of this, are to us the heralds. And mark how consistent the sequel of the tale is; after the return of age to youth, follows the return of the dead, who are lying in the earth, to life; simultaneously with the reversal of the world the wheel of their generation has been turned back, and they are put together and rise and live in the opposite order, unless God has carried any of them away to some other lot. According to this tradition they of necessity sprang from the arth and have the name of earth-born, and so the above legend clings to them.

SOCRATES:

Certainly that is quite consistent with what has preceded; but tell me, was the life which you said existed in the reign of Cronos in that cycle of the world, or in this? For the change in the course of the stars and the sun must have occurred in both.

If we imagine the cycles described by Plato's stranger as repeating endlessly, we have an oscillating model of the universe that is surprisingly similar to Gold's model, as well as to the eternal recurrence doctrines of certain Eastern religions. Let us embroider it with more conjectures—first, the notion that every black hole is associated with a white hole from which all the matter and energy eaten by the black hole gush forth. The two holes are joined by an "Einstein-Rosen bridge," or what John Wheeler calls a wormhole. Perhaps the centers of quasars, and those quasarlike galaxies called Seyfert galaxies, are such white holes. If so, the final entropy death of our universe is being delayed by matter constantly being recycled through black and white holes. At the final collapse of the universe, everything will disappear into a gigantic black hole. Will this be followed by the explosion of a white hole that is the big bang of the next cycle? Are we living in a universe of what we call matter that is the leftover antimatter of the collapsing universe that preceded us?

Each backward-moving cycle in this oscillating model can be interpreted in two ways. If we assume determinism, the second cycle could simply repeat what happened in the previous cycle, but in reverse order. On the other hand, the time-reversed universe, like the faustian universe or any time-reversed world situated somewhere "out there," need not be deterministic at all. It could go backward with an entirely different history. In the absence of any other universe with which it can interact, intelligent beings in the "backward" universe would find themselves moving forward in time in a perfectly normal manner.

There seems to be nothing wrong with the first interpretation except that it seems dull and pointless for history to keep repeating itself in alternate time directions. To the gods it would be like our reading *Finnegans Wake* to the end, then reading it backward to "riverrun," then forward again and repeating this forever, or like watching a motion picture alternately projected forward and backward. But note: we need outside observers in hypertime to give a meaning to "forward" and "backward." If there are no such observers, we might just as well speak of *one* single cycle that never repeats.

The spectacle becomes less boring when the cycles are not identical in their history. Nothing in Plato's vision, or in Gold's, requires each "riverrun" to repeat its history exactly. In each cycle, intelligent beings will find the world just as we do, moving from an immutable past into an unpredictable future. The future, as William James so passionately argued, could be filled with genuine surprises that not even gods could anticipate. It would certainly make the movies more exciting for them if they didn't know how each picture ended.

Let us enlarge this vision even more. Existence contains an infinity of universes, each alternately expanding and contracting, each with genuine "futures" that do not "exist" until they happen. Intelligent creatures in any cycle of any universe deem themselves living forward in time. If there is no interaction of any sort between these worlds, it is hard to see how logical contradictions can arise to make nonsense of such a vision.

A curious thought now arises. What is to prevent radiation from the expanding half of the cycle from continuing on its way and entering into the contracting half? Davies reports in his book that Wheeler has conjectured that there is a gradual "turning of the tide," when the universe slows to a halt and starts to move the other way. If so, then near the end of the expanding cycle one might begin to see a portion of this radiation coming back to us in a blurred, diffuse state. If the direction of time reverses for the contracting phase, this would be (as Davies puts it) "a search for electromagnetic microwaves from the future." At least one experiment, he tells us, has actually been made to look for such microwaves, but it failed to detect them.

Even stranger situations arise in thought experiments in which we imagine individual persons or individual particles going backward in time while the rest of the universe goes forward. We will consider some of them in the next chapter.

29. TIME–REVERSED PERSONS AND PARTICLES

Plato's younger contemporary, the Greek historian Theopompus of Chios, wrote about a certain fruit that, when eaten, would start a person growing younger. This is not, of course, the same thing as a complete reversal of a person's time, because he still thinks, talks, and acts in the normal way.

Several science-fiction stories have been about individuals who grew backward in this way, including one whimsical tale, "The Curious Case of Benjamin Button," by (of all people) F. Scott Fitzgerald. Benjamin is born in 1860 as a seventy-year-old man with white hair and a long beard. He grows backward at a normal rate, enters kindergarten at sixty-five, goes through a normal schooling, and marries at fifty. Thirty years later, age twenty, he enters Harvard and is graduated in 1914 when he is sixteen. He joins the army at the outset of World War I and is immediately made a brigadier general because, as a biologically older man, he had been a lieutenant colonel during the Spanish-American War. But when he shows up at the army base as a small boy he is packed off for home. He continues to grow younger until he cannot walk or talk. "Then it was all dark," reads the story's last sentence, "and his white crib and the dim faces that moved above him, and the warm sweet aroma of the milk, faded out altogether from his mind."

Aside from his backward growth, Mr. Button lives normally in forward-moving time. A better description of a situation in which a person's time arrow points backward is found in Lewis Carroll's novel *Sylvie and Bruno Concluded.* The German Professor hands the narrator an Outlandish Watch with a "reversal peg" that causes him and his nearby environment to run backward for hours. There is an amusing description of a time-reversed dinner at which "an empty fork is raised to the lips: there it receives a neatly cut piece of mutton, and swiftly conveys it to the plate, where it instantly attaches itself to the mutton already there." The scene is not consistent. The order of the dinner-table remarks is backward, but the words are spoken in a forward time direction.

The same inconsistency applies to *Happy End,* a Czechoslovakian movie in which the action runs backward, as well as the sequences of remarks, but not the word sequences. For example:

Wife: "Such sad looking fish."

Lover: "You are too, my dear."

Wife: "The weather is beautiful."

Time magazine concluded its review (June 28, 1968) with: "Much too hour an is it of minutes 73 but."

If we try to imagine an individual whose entire bodily and mental processes are reversed, while the rest of the world remains the same, we run into the worst kind of difficulties. For one thing, he could not pass through his previous life experiences backward, because those experiences are bound up with his external world. Since that world is still moving forward, none of his past experiences can be duplicated. Would we see him go into a mad death dance, like an automaton whose motor has been reversed? Would he, from his point of view, find himself still thinking forward in a world that seemed to be going backward? If so, he would be unable to see or hear anything because all light and sound waves in the outside world would be moving toward their point of origin.

We seem to encounter nothing but logical nonsense when we try to apply a reversed time arrow to an individual mind. Is it possible, however, on the microlevel of quantum theory to speak sensibly about a particle moving backward in time? It is. In 1965 Richard P. Feynman, whom we encountered in chapter 22, shared the Nobel Prize in physics for his revolutionary contributions to quantum mechanics. In Feynman's "space-time view," as it is called, with its ingenious "Feynman graphs," antiparticles are treated as particles moving backward in time for a fraction of a microsecond.

When there is pair-creation of an electron and its antiparticle, the positron (positively charged electron) is extremely short-lived. It immediately collides with another electron, both are annihilated, and off goes a gamma ray. In Feynman's theory there is only *one* particle, the electron (see Figure 68, left). What we observe as a positron is merely the electron moving momentarily backward in time. Because our time, in which we watch the event, runs uniformly forward, we see the time-reversed electron as a positron. We think the positron vanishes when it hits another electron, but the vanishing is just the spot in time

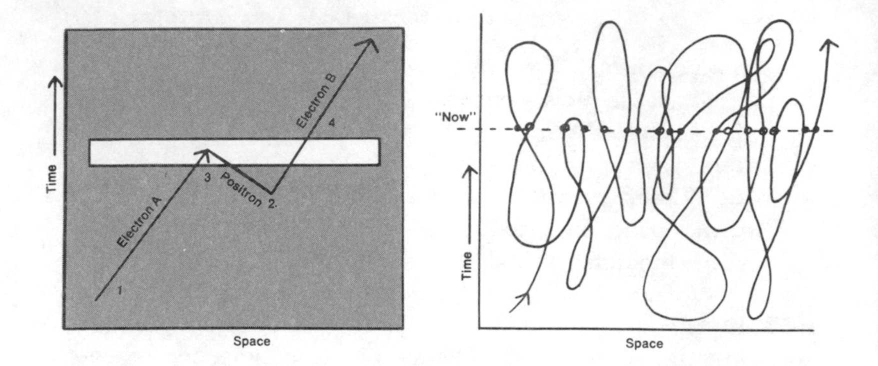

Figure 68. Feynman graphs. The graph shown at the left in a simplified form devised by Banesh Hoffman of Queens College shows how an antiparticle can be considered a particle moving backward in time. The graph is viewed through a horizontal slot in a sheet of cardboard (gray) that is moved slowly up across the graph. Looking through the slot, one sees events as they appear in our forward-looking "now." Electron *A* moves to the right (1), an electron-positron pair is created (2), the positron and electron *A* are mutually annihilated (3) and electron *B* continues on to the right (4). From a timeless point of view (without the slotted cardboard), however, it can be seen that there is only one particle: an electron that goes forward in time, backward, and then forward again. Richard P. Feynman's approach stemmed from a whimsical suggestion by John A. Wheeler of Princeton University: a single particle, tracing a "world line" through space and time (*right*), could create all the world's electrons (*black dots*) and positrons (*white*).

where the electron entered the past. The electron executes a tiny zigzag dance in space-time, hopping into the past just long enough for us to see its path in a bubble chamber and interpret it as the path of a positron moving forward in time.

Feynman got his basic idea when he was a graduate student at Princeton, from a telephone conversation with his physics professor, John A. Wheeler. In his Nobel-Prize acceptance speech Feynman told the story this way:

"Feynman," said Wheeler, "I know why all electrons have the same charge and the same mass."

"Why?" asked Feynman.

"Because," said Wheeler, "they are all the *same* electron!"

Wheeler went on to explain on the telephone the stupendous vision that had come to him. In relativity theory physicists use what are called Minkowski graphs for showing the movements of objects through space-time. The path of an object on such a graph is called its world line. Wheeler imagined one electron, weaving back and forth in space-time, tracing out a single world line. The world line would form an incredible knot, like a monstrous ball of tangled twine with billions on billions of crossings, the "string" filling the entire cosmos in one blinding, timeless instant. If we take a cross-section through cosmic space-time, cutting at right angles to the time axis, we get a picture of 3-space at one instant of time. This three-dimensional cross-section moves forward along the time axis, and it is on this moving section of "now" that the events of the world execute their dance. On this cross-section the world line of the electron, the incredible knot, would be broken up into billions on billions of dancing points, each corresponding to a spot where the electron knot was cut. If the cross-section cuts the world line at a spot where the particle is moving forward in time, the spot is an electron. If it cuts the world line at a spot where the particle is moving backward in time, the spot is a positron. All the electrons and positrons in the cosmos are, in Wheeler's fantastic vision, cross-sections of the knotted path of this single particle. Since they are all sections of the same world line, naturally they will all have identical masses and strengths of charge. Their positive and negative charges are no more than indications of the time direction in which the particle at that instant was weaving its way through space-time.

There is an enormous catch to all of this. The number of electrons and positrons in the universe would have to be equal. You can see this by drawing on a sheet of paper a two-dimensional analogue of Wheeler's vision. Simply trace a single line over the page to make a tangled knot (see Figure 68, right). Draw a straight line through it. The straight line represents a one-dimensional cross-section at one instant in time through a 2-space world (one space axis and one time axis). At points where the knot crosses the straight line, moving up in the direction of time's arrow, it produces an electron. Where it crosses the line going the opposite way it produces a positron. It is easy to see that the number of electrons and positrons must be equal or have at most a difference of one. That is why, when Wheeler had described his vision, Feynman immediately said: "But, Professor, there aren't as many positrons as electrons." "Well," countered Wheeler, "maybe they are hidden in the protons or something."

Wheeler was not proposing a serious theory, but the suggestion that a positron could be interpreted as an electron moving temporarily backward in time caught Feynman's fancy, and he found that the interpretation could be handled mathematically in a way that was entirely consistent with logic and all the laws of quantum theory. It became a cornerstone in his famous space-time view, which he completed eight years later. The theory is equivalent to traditional views, but the zigzag dance of Feynman's particles provided a new way of handling certain calculations and greatly simplifying them. Does this mean that the positron is "really" an electron moving backward in time? No, that is only one physical interpretation of the "Feynman graphs"; other interpretations, just as valid, do not speak of time reversals.

A new light, however, has been cast on the situation by the recent experiments that suggest a mysterious interlocking of charge, parity, and time. The zigzag dance of Feynman's particles, as they jump back and forth in time, no longer seems as outlandish a physical interpretation as it once did. If antiparticles observed on earth can be viewed, without contradiction, as particles moving backward in time, we can easily imagine a universe exactly like ours except that all three asymmetries—charge, parity, and time—are reversed with respect to our world. If it turns out that charge is, in some as-yet-unknown sense, a left-right reversal, then the antiworld is simply a world that is reversed with respect to space and time.

In chapter 17 we recalled Kant's perplexity over the fact that a pair of enantiomorphic solid objects can be exactly alike in all respects except that you cannot turn one around and fit it to the other. We saw how this becomes less mysterious once we realize that the two objects are identical when embedded in a higher space—that their difference is only an illusion that arises when they are trapped in 3-space with opposite handedness.

Physicists today are less perplexed by the phenomenon of spatial asymmetry on the microlevel than they are perplexed by the one-wayness in time of certain microevents. Feynman's vision provides a startling way out that is simply an extension to time of the same trick we used in chapter 17 to clear up Kant's perplexity. Our world and the antiworld could be identical in the same way that right and left hands are identical, only now we have to make two imaginary leaps instead of one—a leap into a higher space and a leap into a higher time. We who are trapped in three dimensions of space and one of time see the two worlds as mirror images of one another and moving in opposite time directions. A hypermind in a higher space-time might see our world and an antiworld as identical.

We have already observed (in a note for chapter 17) that Vladimir Nabokov's novel *Ada* has an antiworld for its setting. In 1974 Nabokov published a marvelous shorter novel, *Look at the Harlequins!*, in which questions about the symmetries of space and time are so essential to the plot that I like to think that the book was influenced by Nabokov's reading of the first edition of this book. What I call the Ozma problem is explicitly described by Iris, the narrator's first wife, as a "stupid philosophical riddle." The narrator, however, does not consider it stupid. On the contrary, he suffers from a peculiar pathology that is his lifelong torment: he cannot, in his mind, imagine how to turn himself around so that left becomes right. Here is how he describes his ailment:

> In actual, physical life I can turn as simply and swiftly as anyone. But mentally, with my eyes closed and my body immobile, I am unable to switch from one direction to the other. Some swivel cell in my brain does not work. I can cheat, of course, by setting aside the mental snapshot of one vista and leisurely selecting the opposite view for my walk back to my starting point. But if I do not cheat, some kind of atrocious obstacle, which would drive me mad if I persevered, prevents me from imagining the twist which transforms one direction into another, directly opposite. I am crushed, I am carrying the whole world on my back in the process of trying to visualize my turning around and making myself see in terms of "right" what I saw in terms of "left" and vice versa.

At the end of the novel the narrator absent-mindedly walks off a parapet at the edge of a village in Catapult, California. He had been unable, when he approached the edge, to swivel himself around. "To make that movement would mean rolling the world around on its axis and that was as impossible as traveling back physically from the present moment to the previous one."

It is this contrast between the unalterable direction of time and our freedom to turn around in space that is at the heart of Nabokov's novel. The narrator's "morbid mistake is quite simple. He has confused direction and duration. He speaks of space but he means time." It is only the past that we cannot turn and reenter. It belongs to what Thomas Wolfe called "the done, the indestructible fabric . . . the strange finality of dark time." Nabokov's narrator survives his accident but, like T. S. Eliot writing in *Ash Wednesday*, he "cannot hope to turn again."

The ability of physicists to "twirl time" (I quote now from the last paragraph of *Look at the Harlequins!*) is like one of those neat formulas that they scribble on a blackboard "until the next chap snatches the chalk." There is no way we can escape from time's one-wayness, to avoid that final "dropping off," that "dying away," when we are catapulted altogether out of space, out of time.

30. EPILOGUE

After the heady fantasies of the last two chapters it is time to get back to more mundane matters; to try to summarize more realistically the present situation in physics with respect to spatial and temporal asymmetry.

Universes are not, as Charles Peirce once said, as plentiful as blackberries. We exist in the only one we know, and it seems likely that science has made only the faintest beginning in penetrating its subtle laws. The few basic laws we know have a fantastically high degree of symmetry with respect to both space and time.

Recent experiments have shown that on the microlevel, in certain weak interactions, there are violations of the three great symmetries of handedness, charge, and time. Why this happens is not clear. There are many conjectures, but little evidence to support any of them.

Does the universe contain galaxies of antimatter? Astrophysicists have as yet no way of knowing, but most of them believe the answer is no. In standard big bang models all the matter of the universe is identical with the matter of our galaxy, and going the same way with respect to all arrows of time.

There is not the slightest evidence for the existence of another universe where charge, parity, or time (or any two or all three) is reversed with respect to our world. Indeed, there is no evidence that *any* other universe exists.

We know our universe is expanding. The amount of matter it contains is believed to be insufficient to cause space-time to close on itself. If this is true, the universe is doomed to go on expanding forever. Even if it turns out that the universe has enough matter (hidden away where astronomers have not yet found it) to close space-time and reverse the expansion, there is no reason to suppose that this reversal would have any effect on charge, parity, or the microlevel direction of time. Nor is there any evidence that if the universe began to contract, the macroarrows of entropy, radiation, and consciousness would obligingly turn around. It would probably be the same old universe, just getting smaller instead of larger, and heading for a singular destiny about which there is now a great deal of controversy.

The Ozma problem appears to be solved. As we have seen, if a time-reversed galaxy exists, we cannot communicate with anyone in it. If communication can be established at all, we and they can perform tests that will provide a common understanding of left and right.

Paintings and statues do not have to be left–right symmetric to be beautiful. Why should the universe be considered less beautiful because it is lopsided in various ways? Scientists who are offended by such symmetry violations do not really need an antiworld to restore aesthetic satisfaction. They certainly do not need to assume that our world would turn into an antiworld if it began contracting.

A much simpler way to obtain aesthetic satisfaction was considered briefly in chapter 24. Just before the big bang, everything may indeed have been completely symmetrical with respect to charge, parity, and even time. When the explosion occurred, the laws of probability may have introduced strong deviations from symmetry. We have seen how the handedness of living molecules may be an accidental fact that could just as easily, in earth's primeval seas, have gone the other way. So with the lopsidedness of matter. Symmetry on the microlevel is restored by the simple hypothesis that, in the first few microseconds of the universe's history, things could *just as easily have gone the other way.* A statue by Michelangelo is asymmetric. Why be dismayed because its mirror-image twin doesn't exist? We know that Michelangelo could have modeled the statue the other way, and we can even see how it would have looked by viewing the statue in a mirror.

A round bowl of soup has a high degree of radial symmetry. So does a small cork ball. What happens when you drop the ball on the exact center of the soup? Although the liquid's forces are symmetrical, and the laws of probability have no preferred direction, the cork ball will not remain at the center. Its position there is unstable. The ball will float to the bowl's rim and the beautiful radial symmetry of the system is destroyed. In similar fashion the slightest statistical fluctuation of the hypothetical and symmetrical quark soup, when it exploded into our cosmos, could have created an imbalance that produced matter instead of antimatter. We are back to the ancient Yin-Yang symbol. It is asymmetric, but we can draw it either way.

From this point of view, the matter that survived the primeval fireball acquired by accident its present spatial twist, charge, and the arrow of time we now observe in weak interactions. We can easily conceive of a universe in which these three symmetries all go the other way. We cannot conceive of a universe, arising from a big bang, without a unidirectional macrotime.

Now comes a subtler notion. In the absence of any other universe for comparison, and leaving aside the notion of our universe being observed by hyperbeings, a CPT-reversed universe is as meaningless as a universe turned upside down. Even if we imagine our universe observed by intelligences in a higher space-time, a CPT-reversal is equally meaningless. It is comparable to our turning over a cardboard cutout of the letter *R*. A hyperbeing simply twirls our universe around in space and time, and looks at it from another point of view.

One of the greatest lessons that can be learned from the history of science is one of humility. Science may indeed be steadily learning more about the structure of the world, but surely what is known is exceedingly small in relation to what is unknown. There is no scientific theory today, not even a law, that may not be modified or discarded tomorrow. "The great invariant principles of nature," wrote Philip Morrison in "The Overthrow of Parity" (*Scientific American,* April 1957), "may be relied upon within the domains of their application, but they are not *a priori* self-evident or necessarily of universal application. It is worthwhile to test to higher and higher precision the great fundamental principles. . . . We have entered an exhilarating time."

One of the most exhilarating prospects at the moment, which some physicists believe is almost upon us, is the construction of a deep theory of particles that will explain, in some elegant mathematical way, why all the particles are what they are. Abraham Pais, writing on "Particles" (*Physics Today*, May 1968) described particle physics as in a state "not unlike the one in a symphony hall a while before the start of the concert. On the podium one will see some but not yet all of the musicians. They are tuning up. Short brilliant passages are heard on some instruments; improvisations elsewhere; some wrong notes too. There is a sense of anticipation for the moment when the symphony starts."

If we could now hear a few strains of the great new symphony, the music might well strike us as insane. Freeman Dyson, in an article from which we quoted in chapter 22 ("Innovation in Physics," *Scientific American*, September 1958), recalled that in 1958 Werner Heisenberg and Wolfgang Pauli put forward an unorthodox theory of particles that would explain the violations of parity in weak interactions. Pauli was lecturing in New York on these new ideas to a group of scientists that included Niels Bohr. In the discussion that followed the talk, younger scientists were sharply critical of Pauli.

Bohr rose to speak. "We are all agreed," he said to Pauli, "that your theory is crazy. The question which divides us is whether it is crazy enough to have a chance of being correct. My own feeling is that it is not crazy enough."

Dyson commented in his article:

"The objection that they are not crazy enough applies to all the attempts which have so far been launched at a radically new theory of elementary particles. It applies especially to crackpots. Most of the crackpot papers which are submitted to *The Physical Review* are rejected, not because it is impossible to understand them, but because it is possible. Those which are impossible to understand are usually published. When the great innovation appears, it will almost certainly be in a muddled, incomplete and confusing form. To the discoverer himself it will be only half-understood; to everybody else it will be a mystery. For any speculation which does not at first glance look crazy, there is no hope."

And to Dyson's wise words I would like to add (though I cannot claim to be a scientist): After the crazy theory has been refined until it no longer seems crazy but simple and almost inevitable, and the apparent disorder of the particles has given way to a beautiful order, the very success of the theory will unlock doors leading to a deeper level of dishevelment.

I do not belong to that incredible school of thought which believes that science will someday discover everything. Such a view strikes me as an expression of simple-minded arrogance, and I am at a loss to know how to converse with anyone who holds it. Surely, to adapt a well-known metaphor of William James, there are truths about existence as far beyond the range of our minds as Dublin is beyond the mind of a fish swimming in the river Liffey.

"A man is a small thing," remarks King Karnos, in Lord Dunsany's play *The Laughter of the Gods,* "and the night is very large and full of wonders."

ANSWERS TO EXERCISES

CHAPTER 2

1. The asymmetric letters are F, G, J, L, N, P, R, S, Z.

CHAPTER 3

2. A cube has nine planes of symmetry. Three are parallel to pairs of opposite faces, six pass through pairs of opposite edges.

3. The right-handedness of screw and bolt threads reflects the dominant right-handedness of the human race. If you hold a screwdriver in your right hand, a stronger twisting force can be exerted clockwise than counterclockwise because it brings the powerful biceps muscle of the arm into play. In addition, the fleshy base of the right thumb applies greater frictional resistance to a screwdriver handle when it is twisted clockwise. (I am indebted to Dr. Harvey P.

Kopell, of the New York University School of Medicine, for calling both points to my attention.)

4. All the objects except the bowling ball are fundamentally asymmetric. The monkey wrench has an asymmetric worm gear for opening and closing its jaws. Some three-hole bowling balls have a symmetric pattern. Of course all two-hole bowling balls are symmetrical.

CHAPTER 4

5. If you turn your head you will see "mud" on the wall behind you. On the wall in the mirror you will see "bum."

This exercise is isomorphic with a common experience of car drivers. Suppose a transparent strip with the name of a university is stuck to the inside of the car's rear window so that anyone behind the car can read it correctly. If the driver of the car turns his head around he will see the words reversed, but in the rear-view mirror he reads them normally.

6. The top faces of the dice, from top down are: 5, 3, 1.

CHAPTER 10

7. Men and women are in the habit of buttoning only their *own* coats. In buttoning a double-breasted trenchcoat on someone else, a man would be more likely to carry the side in his left hand over the side in his right and so produce a mirror reflection of the way he buttons the coat on himself.

CHAPTER 11

8. Each of the cube's four threefold axes of symmetry passes through two diametrically opposite corners.

9. The rhombohedron has three planes of symmetry, each passing through a pair of opposite edges.

CHAPTER 12

10. The model of the grain alcohol molecule has one plane of symmetry. It passes through the centers of the cherry and marshmallow, cutting the structure in half.

CHAPTER 17

11. The word *TUO*, viewed from the other side of the glass door, is *OUT*.

12. Yes, all one-sided surfaces are "nonorientable." This means that an asymmetric figure on the surface can be reversed in handedness by sliding it a certain way around the surface and back to where it started.

CHAPTER 19

13. The procedure does nothing more than create a bar magnet in front of one, with its north pole on the right. Until the meaning of *left* and *right* is communicated, the Venusian would not know which end of the bar is the one we call north.

CHAPTER 20

14. It is not possible to bring the glasses either all upright or all upside down. At the start, an odd number of glasses are brim up. If you reverse two glasses that are brim down, you increase the number of up glasses by 2, so the total number of up glasses remains odd. If you turn two up glasses, you decrease the number of up glasses by 2, so the total also remains odd. And if you turn two glasses that face opposite ways, you take away one up glass and add one up glass, so the number of up glasses does not change. It is therefore impossible, by turning the glasses in pairs, to change the number of up glasses to even. Since there are six glasses, an even number, you can never bring all the glasses face up. The same argument proves that they cannot all be brought face down.

CHAPTER 21

15. The statement that Dr. Teller and Dr. Anti-Teller shook "right hands" can be interpreted in four different ways:

1. Each extends what *he* regards as his right hand. (A photograph of the event would appear to us as if Dr. Teller, with his right hand, clasped the left hand of Dr. Anti-Teller.)

2. Each extends what we would call his right hand. On this interpretation, the poem is written from our point of view.

3. Each extends what would be called, on antiearth, a right hand. On this interpretation, the poem is written from an anti point of view, making the ensuing explosion an anticlimax.

4. Each extends what he considers his left hand, but which the other regards as a right hand. This is the most farfetched of the four interpretations.

CHAPTER 22

16. Observation of the experiment would have to include observation of details that indicate the direction of current flow through the wires that coil around the electromagnets. This direction, together with the handedness of the wire helices, would establish which pole of the electromagnet is the one we traditionally label south. If the majority of electrons were emitted from the corresponding south end of the cobalt nuclei, the motion picture would be unreversed. If they were emitted from the north end, it would prove that the film had been reversed.

17. The 3-space analog of the Yin-Yang symbol is a baseball. It is left–right symmetrical. Piet Hein, the Danish poet-inventor, has suggested that a better analog would be a sphere's surface divided into *three* identical regions with an overall pattern of asymmetry. Imagine a cubical shell made of an elastic material. Paint its left and back faces red, its top and right faces white, its front and bottom faces blue, then inflate the cube until it becomes a sphere. The three colored regions remain congruent, like the Yin and Yang, and the overall pattern is asymmetric. Piet Hein proposes calling the regions Yin, Yang, and Lee.

INDEX